Praise for
ERIC BRENDE and *Better Off*

"Deftly steering clear of dogma, never sounding like a sanctimonious scold, Eric Brende makes a persuasive case that most of us would enjoy life more by radically minimizing our reliance on modern technology. *Better Off* is a buoyant, thought-provoking, and very entertaining read."　　　　　—Jon Krakauer, author of *Into the Wild, Into Thin Air,* and *Under the Banner of Heaven*

"A journey into another world . . . a sobering assessment of a society that drives to the gym to get exercise."　　　　　—*Wall Street Journal*

"This calm and unjudgmental account of living a little more lightly on the earth will do readers more good than a thousand 'self-help' books crowding bestseller lists. It will make you think about your own life more than you've thought about it for years, and for that service we can be deeply grateful to the talented Eric Brende."
　　　　　—Bill McKibben, author of *The End of Nature* and *Enough: Staying Human in an Engineered Age*

"I recommend spending a few hours of your valuable time reading *Better Off*. It will give you a new perspective on your life and will be time well spent. . . . It is a good story well told. The fact that there is conviction of thought behind the words makes it even easier to read and relish. If *Better Off* doesn't make you examine your life, you may be too far gone already."　　　　　—Dan O'Brien, *Boston Globe*

"Gracefully written and inspirational."　　　　　—*Hartford Courant*

"For those who want to get away from modern life and technology altogether, Eric Brende presents his own interesting and evocative personal experience, demonstrating that it is possible to pursue Thoreau's ideals today, and perhaps emerge the richer for it."

—Michael Korda, author of *Country Matters*

"I more than read this book . . . I lived with it. I read it slowly, intentionally, as one does a good novel, because I wanted it still to be there for me. The book most reminded me of Robert Pirsig's *Zen and the Art of Motorcycle Maintenance* in that the journey is a vehicle for a wide range of observations, musings, and philosophizing. . . . The writing is finely crafted and perfectly paced—a pleasure to read, and the message is clear and compelling without being preachy or self-righteous. Indeed, the tone matches the theme, a blend and balance of insights and humility. This book works brilliantly."

—David F. Nobel, author of *The Religion of Technology* and *Forces of Production: A Social History of Industrial Automation*

"Offers fascinating insights into the workings of a Mennonite community. . . . May stimulate readers to think about the role modern technology plays in their own lives." —*Christian Science Monitor*

"A fascinating narrative. . . . Brende's observations induce rather than hector readers into appreciating the neighborly benefits of renouncing telephones, processed food, and cars. This memorable personal story will warm the heart of anyone dreaming about an alternative close-to-the-land lifestyle." —*Booklist*

"[Brende's] gentle case for simple living will easily resonate with the converted and may inspire skeptics to grapple more intimately with the issue." —*Publishers Weekly*

"Brende's lofty and humbling endeavor raises some interesting questions."
—*Library Journal*

"[Brende's] glimpse of low-tech living will leave you more mindful of the price of every modern convenience." —*Body & Soul*

"[Brende] is an inspiration to anyone seeking to escape the tyranny of twenty-first-century machinery and smash the mechanical paraphernalia of our civilization to smithereens." —*Daily Telegraph* (London)

"Brende's book doesn't merely make a case, it tells a gripping story about turning your back on technology, or (at a deeper level) acting on what you believe. The book is hard to put down because the story is so good, the life it describes is so wistfully attractive, and the prose has the soothing, mesmerizing rhythm of a forest stream. You want to read *Better Off* all day long and get lost in it—and dream about a different, better life; the book is a vicarious Utopia. It will make you sad, and you'll remember it for a long time."
—David Gerlernter, author of *Surviving the Unabomber* and *Mirror Worlds: Or the Day Software Puts the Universe in a Shoebox: How It Will Happen and What It Will Mean*

better
off

BRODART, CO.   Cat. No. 23-221-003

better
off

Flipping the Switch on Technology

ERIC BRENDE

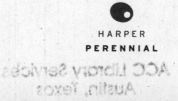

HARPER

PERENNIAL

HARPER ● PERENNIAL

Portions of this book appeared in another form in *Technology Review/New York Times Syndicate, Mother Earth News, New Oxford Review,* and *Caelum et Terra.*

A hardcover edition of this book was published in 2004 by HarperCollins Publishers.

P.S.™ is a trademark of HarperCollins Publishers.

HarperCollins books may be purchased for educational, business, or sales promotional use. For information please write: Special Markets Department, HarperCollins Publishers, 10 East 53rd Street, New York, NY 10022.

FIRST HARPER PERENNIAL EDITION PUBLISHED 2005.

Designed by Christine Weathersbee

Map courtesy of Justin P. Leland

The Library of Congress has catalogued the hardcover edition as follows:
    Brende, Eric.
        Better off : flipping the switch on technology / Eric Brende.—1st ed.
        p. cm.
    ISBN 0-06-057004-0 (acid free paper)
        1. Technology—social aspects.   2. Technology—Psychological aspects.
    3. Simplicity.   4. Farm life.   5. Brende, Eric.   I. Title.
    HM846.B74 2004   2003067567
    303.48'3—dc22

ISBN-10: 0-06-057005-9 (pbk.)
ISBN-13: 978-0-06-057005-7 (pbk.)

05 06 07 08 09 ❖/RRD 10 9 8 7 6 5 4 3

To Mary

(and in repose, John Senior)

Acknowledgments

I cannot begin to thank the many people who, at various times and at various stages of the manuscript leading up to this book, generously chipped in their comments and suggestions regarding anything from the phrasing of a single line to the thrust of the entire narrative: Paul Chu, Bette Denich, Wade Roush, Dave Clemenson, Mike Stubbs, Mark Edmundson, George O'Har, Dennis Quinn, Daniel Nichols, Mary Byrne, Steve Faulkner, Beth Huber, Will Hoyt, Chuck Presberg, Sherry Hausman, Mary the librarian, Pat Wendleton, Becky Penrod, Sandra and Ralph Meredith, Michelle of Writer's Inc., Michael Romick, Nat and Jane Griffin, Dickson Beall, Jordan House, James Howard Kunstler, Alan Whitney, Chris Paciorek, Mom, Larry, Jackie, and Dad (whose insight contributed greatly to the title, and thanks for their title suggestions as well to my other family members, Alicia, Cary, Mark, Jenny, Kareen, Jeff, Terri, Bob, and Tommy); also Eda Kranakis, Pauline Maier, and Merritt Roe Smith, Uday Mehta, and Betty Cole (who, though never reading any of this, provided indispensable background information and encouragement). Thanks also to Steve and Cathy at Community Copy and the generous Hannegan family. Kudos to Justin Leland for the map. Forgive me if I have inadvertently omitted anyone.

I must acknowledge a special debt to those whose substantial encouragement and advice at critical moments were especially vital: Leo Marx; David Noble; Dick Sclove; Homer White; Jeffrey Ruckman; Randy Testa; Paula and Jim Nedved; my indefatigable agent, John Ware; and my editor at HarperCollins, Alison Callahan.

And most of all, hats off to the "Minimites," without whom this book would not have been possible.

Explanatory Note

What follows in these pages is the story of an unusual adventure taken by my wife and me as an adjunct to my graduate studies at M.I.T. Because of the highly personal nature of much of the material and the low tolerance for publicity of the extraordinary people we lived with, their names have been changed, minor details altered, and the location kept secret. Nonetheless, although certain liberties have been taken with chronology and background circumstance, everything described in this book actually took place.

Readers have some options in how they choose to proceed. The story can be read the way stories usually are, that is, as entertainment (I hope riveting), or as food for thought on the broader human condition (I hope stimulating), or even in this case as a real-life model for practical action (I hope instructive).

For those inclined there is still another way to read the book, and that is as a hands on experiment attempting to establish, among other things, the answer to a key question about technology—namely, How much is enough?

Contents

Cast of Neighbors

(in order of appearance)

MR. MILLER—our landlord and mentor
MRS. MILLER—his wife
AMOS, CALEB, JUDY, ELLIS, IRMA, JED, and
    HAROLD—their at-home children
NATE and CAROL JONES—fellow escapees from
    the technological world
GIDEON STOLTZFUS—builder of the water mill
    and horse-powered turnstile
ALVIN STOLTZFUS—collaborator with Gideon on
    the water mill
SYLVAN—Mr. Miller's son-in-law, collaborator
    on my sorghum crop
BISHOP HENRY—bishop of the community
MINISTER JAMES—skilled preacher
ROB, JOHN, HOWIE, ELI, RED, SIM, ELBERT—crew
    at the barn raising
BILL—novice from the outside world
FRED—participant in a hoeing
EDWARD—Bill's well-educated boss and mentor,
    candidate for the ministry
GRACE—Edward's wife
HARVEY—Mr. Miller's oldest married son and pig
    salesman
GERTIE—his wife
CORNELIUS—a hermit and saw sharpener
EDNA—our midwife
WILBUR—proponent of green manure, candidate
    for the ministry
NAOMI—the midwife's apprentice (Mr. Miller's
    sometimes-away daughter)
MR. BERNHARDT—adventurous would-be motorist

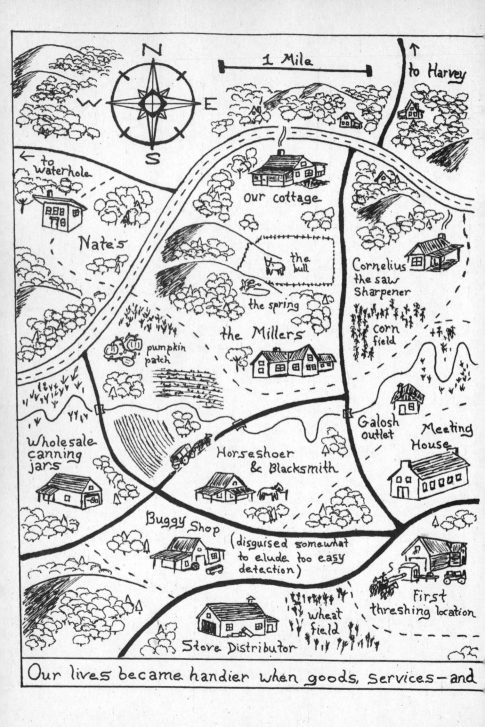

N
W E
S

1 Mile

↑ to Harvey

← to waterhole

our cottage

Nate's

the bull

the spring

Cornelius the saw Sharpener

the Millers

pumpkin patch

corn field

Wholesale canning jars

Horseshoer & Blacksmith

Galosh outlet

Meeting House

Buggy Shop (disguised somewhat to elude too easy detection)

First threshing location

wheat field

Store Distributor

Our lives became handier when goods, services—and

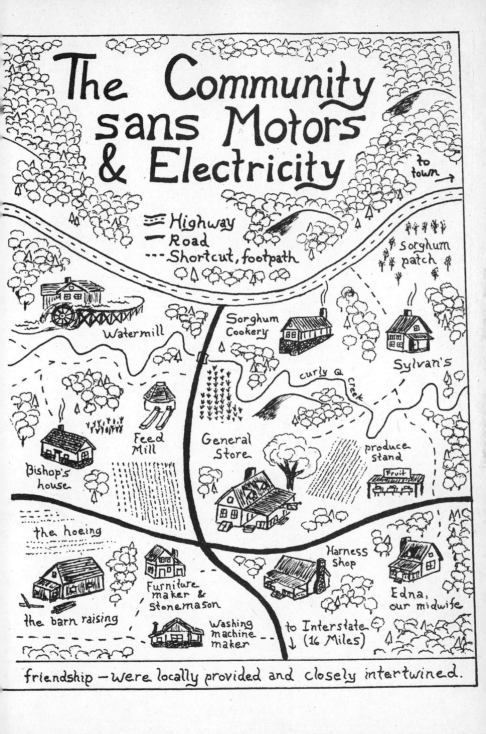

better
off

taking orders

The other day someone I know was running an errand in the outer reaches of a major city. She rarely drives, but for some reason had this day dared to hazard the maze of crowded freeways and interchanges. When rush hour came on, she quickly wearied and pined for a break. A sign for a fast-food place came up, and against her better judgment, she pulled over. Seeing no one at the drive-thru, she slid into the ordering station.

She waited three full minutes before anyone responded, even though hers was the only car in line. Then a male voice came over the microphone:

"Just one moment please."

She waited another sixty seconds.

Finally the voice resumed: "May I take your order?"

"One small French fries please."

"Anything else?"

"No thank you."

"Pull ahead."

At the window a young man who appeared well fed on fries said, "Uh, we're learning to use a new cash register, so it'll be about a ten- to fifteen-minute wait."

"Ten to fifteen minutes for one order of French fries?"

There was a pause as the young man took in my friend's bewilderment.

"Uh, I see what you mean. But I don't know how to do it. Why don't you pull ahead to the next window."

At the next window, a slightly older female, who might have been the manager, again fielded the request. At first she gave a blank stare. Then a light went off in her head. Taking out pencil and paper, she did some arithmetic and, without any further ado, handed over the fries.

My friend drove away puzzled. Now fully revived, she began to unravel what had happened. It appeared to her that the workers were involved in a chain of command with a machine at the top. It told them what the prices were, it added them together, it cued them to their duties. They took their orders from a cash register. Only when pressured by a customer did the manager at last realize that she had the power to act of her own accord and, in a flash of recognition, overrode the mechanical spell.

planting

ONE

seeds of discontent

I used to be as optimistic as anyone about technology. Once asked in grade school to draw a picture of what my home would look like when I grew up, I sketched, in crayon, a transparent hemisphere resting on a single pole and a little flying saucer containing me, my wife, and our many kids about to dock at it. There were exactly eight little heads (besides mine and my wife's) peeking over the rim of the craft, all identical and propagated with the help of a fertility drug.

When I reached my early teens, I never failed to watch an episode of *Star Trek,* and I read almost every piece of science fiction Isaac Asimov wrote. On our family's first cross-country trip, I became ecstatic when we got caught in a traffic jam on the Oakland Bay Bridge. To a midwestern boy, traffic jams were exotic events in which only special people living in modernistic cities took part.

There was always an undertow to my technological infatuation, however, which at first I was loath to acknowledge. On that trip out west, I spent most of the time carsick. A few years later, after returning to the lazy metropolis of Topeka, Kansas, I began to notice anomalies in the mechanical utopia of our modernized household. After we got an automatic dishwasher, the size of the pile of dirty plates on the countertop didn't decrease at all. If anything, it increased. My dad bought one of the first word processors ever made in the hopes of easing the time and effort of writing. He spent so much time with that machine, I almost never saw him again.

In my grade-school years, the neighborhood seemed alive with children out in the street playing stickball and hide-and-seek. But the older I grew, the more deserted the street became—except for the cars, of course, which had multiplied over time and made playing out-of-doors more perilous. After supper even the cars went into hibernation; the only signs of life were the faint glows cast by cathode ray tubes on living-room blinds.

I had always been on the bashful side, so I went more or less the way of the trend, retreating as determinedly as everyone else to the altar of TV. But lest I surrender utterly to The Void, I applied myself diligently at the piano, practicing several hours a day. To survive socially in a place dominated by the automobile, of course, you had to drive; so I also made an attempt to earn money to buy a car by working at McDonald's. But soon I saw the futility and the irony: in a town whose borders motor vehicles had pushed to the horizons (with a population of 120,000, Topeka covered 50 square miles), the only sensible way to get to my job was by automobile. Until I could afford one, I had to bike the six-mile round trip on busy roads with no shoulders or sidewalks, and I arrived dripping wet. Had I stayed on, I calculated that, like the other workers, I would be working mostly in order to pay for my transportation to work.

What had begun as car sickness in boyhood had developed, by adolescence, into a deeper case of cultural indigestion. It was only when I got to college that I began the attempt to put a name to this, but already the symptoms of the malady—burdensome material inconvenience and social isolation—had become too acute to ignore.

Luckily, my musical diligence paid off, and I got into a good university. There it was exciting to meet other people of similar interests who lived within walking distance. I threw out the sheet music and threw myself into the life of the campus. I joined debating groups. I took up rowing. I made new friends. I dabbled in religion. And in my academic pursuits, I tried to gain some understanding of what was going wrong in Oz.

On a hunch, I signed up for a course in the history of technology. It was an eye-opener. The young professor, Eda Kranakis, capably surveyed the development of wind- and water mills, steam engines, and railroads, and tossed in a graphic description of the inhuman working conditions in nineteenth century factories. She related the tragic tale of the British land enclosure movement, inspired by "scientific farming," which uprooted countless laborers from their hereditary commons in the country and flung them into the cities, where they formed an easily exploited labor pool.

As illuminating as the class was, though, it raised more questions than it answered. Hadn't American society moved beyond the barbarities of Dickensian England (or at least hadn't it subcontracted the dirty work to countries like Mexico)? What was technology's role in the present age? Problems hadn't disappeared; they were just different. But the exponents of public policy remained about as starry-eyed as I had been in grade school. Even the leaders of my elite university accorded every latest gizmo a virtual hero's welcome. Appalled by this mindlessness, I engaged in many heated discussions with classmates. And I wrote an extended research paper for Kranakis, describing the unhealthy side effects associated with sedentary stress and the use of ordinary automated devices. Kranakis liked the paper and encouraged me to develop my ideas.

The conviction was growing in me that the besetting problem was our culture's blindness to the distinction between the tool and the automatic machine. Everyone tended to treat them alike, as neutral agents of human intention. But machines clearly were not neutral or inert objects. They were complex fuel-consuming entities with certain definite proclivities and needs. Besides often depriving their users of skills and physical exercise, they created new and artificial demands— for fuel, space, money, and time. These in turn crowded out other important human pursuits, like involvement in family and community, or even the process of thinking itself. The very act of accepting the machine was becoming automatic.

By the time I graduated from college, my original orientation had become completely reversed: once an overawed vassal, I now burned with the desire to rise up and battle the technological dragon that, in my view, held society hostage. I found out about a new course at the Massachusetts Institute of Technology that was intended to provide a critical overview of the social effects of machines on human life. It was called "Science, Technology, and Society." I applied and was accepted.

And I found myself in the very den of the beast. Life was never so ticklish. On the one hand, the fledgling S.T.S. graduate program lured me with the promise of free-ranging discussion of all things technological, pro and con. On the other hand, S.T.S. depended on M.I.T. for its existence. There were certain unspoken limits to the discussion, certain subjects not to be broached. The dragon I hoped to slay held me in its palm. I tried to respect my position, but I was not always as cautious as I might have been. The tongue would slip. The papers I wrote would poke a bit too hard.

The trouble may have begun when I signed up for a political philosophy course taught by a young professor from India named Uday Mehta. He showed how our modern Western legal system favored the development of technology: In the seventeenth century the mathematician René Descartes revolutionized philosophy by reconceiving the human being as a machine. Fifty years later the political theorist John Locke handed the latter rights. Locke, arguably the greatest single influence on the writers of the United States Constitution, expounded a novel notion of "property." By snipping apart the complex social obligations of the past, he created a discrete personal domain of freedom, or individual right. This space he called "property," whether it referred to a person or to a thing the person owned. A hereditary commons available for a mixture of uses and livelihoods had little place in Locke's thinking. His idea was to make the human individual and possessions inviolable. But since property was now synonymous with the person, the two became legally akin. By this innovation the machine got a toehold; it gained stature tantamount to that of its owner, even as the owner descended to the plane of the machine. The machine had acquired citizenship.

I wrote a paper for Mehta suggesting it be revoked.

Historians Merritt Roe Smith and Pauline Maier, in another course, unfolded the practical impact of technology on our society, weaving small-scale studies of traditional communities into the larger fabric of the young American republic. It was interesting to find that technology did not always receive the ecstatic greeting it does today. In the early nineteenth century there were uprisings against its encroaching domination, both here and abroad. Smith pointed out the unrest, leading, for instance, to violence against management at the Harper's Ferry armory after new technology was imposed on skilled gun makers. If there was one moment when the scales tipped irrevocably in favor of machinery in the English-speaking world, however, it was probably back in 1817, when the legendary Ned Ludd and his followers were hanged for vandalizing the power looms that were ruining their livelihoods. At that moment came the fulfillment of Locke's novel definition of rights: destroying a machine became legally tantamount to murder. The plea of self-defense counted for nothing, and opposition to technological "advances" has ever since been informally stigmatized as "Luddism."

I wrote a paper for Smith expressing my sympathies—for Ludd.

Midway through the semester, I made an appointment to speak with the S.T.S. program director about my academic progress. Professor Keniston, lanky and affable, was a psychologist by training. For the time I had known him, he had patted my ego in an avuncular way. The rigors of academic life were taking their toll, and I hoped the meeting would help revive me.

Shortly after I walked in his office, however, I sensed a chill. "Eric," Keniston asked with a frown, "do you really want to eliminate laborsaving devices?" The question came out of the blue. It had nothing to do with why I had come to talk to him.

I was stunned and for a split second couldn't speak. I finally murmured a denial. The director went on.

"I just came back from my summer home in Maine. I was trying to move some heavy rocks in the backyard. That pretty well did me in. Took me three days. And I'm not about to do it again soon. Next time I'll use a backhoe."

I shuffled back to my graduate stall in a daze. Keniston had got me all wrong. It wasn't technology *per se* that I was after so much as an attitude of indiscriminate license, a bias in favor of machines over the interest of human beings. But this was M.I.T., after all. In the future, I'd watch my step.

Still, even Keniston had given manual labor a try, and that was partly why I was in trouble. His misadventure helped me to clarify the purpose of my quest: not to rid the world of technology but to ascertain more carefully how much—or how little—technology was needed. Was there some baseline of minimal machinery needed for human convenience, comfort, and sociability—a line below which physical effort was too demanding and above which machines began to create their own demands? Or if there was no such absolute midpoint, was there perhaps a rule of thumb or a formula for arriving at practical compromise in varied circumstances?

In Keniston's sentiment, too, I divined a hint of the source of the homage (to which I myself had once been prone) that technology widely summoned: the fear that reducing it at all would send us backwards to the way things were "before technology"—to a nasty, brutish struggle for sheer survival. Technology's very success in certain tasks incited a broader dread at its absence. All the more reason to clarify the terms of moderation, the only alternative to limitless increase and blind veneration.

Still, I was at a loss. Having framed the question in this way, I faced another problem. Until now my effort had been waged solely in the realm of ideas. But as Keniston's travail showed, the real test of the matter lay in a concrete demonstration—a real-life experiment. It was probably not something I could carry out at an institute of technology.

The architecture of M.I.T. did little to endear me to its purposes. Most designs seemed to have taken their inspiration from the Pentagon. Newer buildings were as faceless and angular as the older ones. Even the neoclassical centerpiece of campus, the majestically domed edifice at 77 Massachusetts Avenue, somehow brought to my mind a nuclear reactor. Behind the facades the classrooms were sterile and the hallways straight and seemingly unending—one was actually nicknamed the "infinite corridor." As a metaphor for the whole, the description fit—a means without an end.

In one of the few humane recesses of this vast impersonal complex, a remote corner of the S.T.S. student-faculty lounge, I found an old piano. No one ever used the room, so I could run through

Chopin's *Barcarolle* or *Fantasie-Impromptu,* or toss in a boogie-woogie rendition of "The Flight of the Bumble-Bee," without being overheard. The passions flew free, and I savored a sense of sweet melancholy of the Polish master. After a session in this mood-venting chamber, I felt something like wine flowing through my extremities, as if my life-blood were being restored.

Then the compliments came. To my surprise, faculty and staff had begun leaving their doors slightly ajar in order to catch snatches of lyricism wafting through the corridors. Those hard bare walls had one advantage after all: their very emptiness opened up a resounding cavity that could be filled by what they lacked—soul, passion, spontaneity, or at least a musical distillate of these.

Between semesters, when money permitted, I traveled back to Kansas, and on one of these journeys I decided to take the bus. It wasn't any cheaper than the plane, thanks to an airfare war, but I guess I thought if I were ever going to try out more rudimentary technology, I might as well do so now.

It wasn't as bad as I thought it might be. The weather was nice, the scenery beautiful. There was plenty of time to unwind and relax and forget. The only tedious moments were the rest stops . . . until Pittsburgh. There I noticed a man getting on the bus with a full dark beard and wide-brimmed black hat. He looked Amish. I eyed him wistfully—and warily. I had once visited Lancaster County in hopes of finding a patch of human culture unravished by machines. At first I was taken in by the lush cornfields, immaculate white barns, two-hundred-year-old houses, old-fashioned buggies, and traditional costumes. Then I found out something odd: the yeomen had loopholes in their rules. They didn't own cars, but they could *lease* them. Telephones were off limits in the house, but not in booths *outside.* Appliances were verboten if they ran on electricity, but not if on pressurized air, propane, or gasoline. Retired farmers often moved to Florida and congregated in special Amish condominiums. Locally they commuted to work from new subdivisions in hired motor vehicles. They even manned the crews that built the subdivisions. They did almost everything everyone else did, using a substitute device. What was the point?

Admittedly not all of them had gone hog wild on machinery; some upheld the spirit of their rules and perpetuated lively practices of reciprocal aid and traditional agrarian labors. But they were fast becoming a minority. Hence my ambivalence when I spied the black-hatted man.

During a rest stop a few hours later, I noticed him standing beside the bus by himself. He looked a little lonely, so I sidled over to him. He seemed glad for the company. His features were fairly striking— his brows jet-black, his eyes alive and fiery. You could tell English was not his first language from the way he clipped his syllables and sometimes groped for a word. Though he was a bit equivocal about his origins, it seemed safe to assume he was Amish.

But he was not from Lancaster County. This man and his neighbors inhabited an area far from the dense human throngs of the eastern seaboard. Somewhere deep in America's heartland, they lived the life the Lancaster Amish had mostly discarded. They brought in the sheaves over their shoulders and hauled them to their barns by horse-drawn wagon. They husked corn by hand, loaded hay loose, and cut firewood with bucksaws. They got by without electricity, telephones, and motor vehicles. And, he said, they observed a rule prohibiting *all* motors. This was a stricter standard than even the most conservative Amish usually upheld. My curiosity was piqued. Was he Amish or not? He hesitated. "You can say so if you like," he said. He shrugged his shoulders as if, for all he knew, the scythe might well be taken up again by farmers everywhere.

A thought came to me, and my heart began to beat faster. I mustered some will, then made him a proposal. He grew silent. Shortly the twinkle returned to his eye, and he gave me his address.

Once having taken the bus, I was really getting carried away.

It was odd how my journey into the heart of technology detoured so readily to these untechnological yeomen. It was even odder how it led to my discovery of a research assistant.

I had one more semester of classes to complete, and as I did so I continued bicycling back and forth to school from my apartment off-campus. As I became more familiar with the route, a pattern of carelessness caught up with me.

It was a gray and drizzly day, and the car was gray too. It had the right of way, but I didn't see it coming. There was an impact, and the next thing I knew I was on the pavement.

I ended up in the emergency room of the Cambridge City Hospital, and after three hours' waiting, I was told nothing was wrong with me. The doctor was puzzled about the intensity of my pain. He guessed it came from a deep bruise, close to the bone in my upper left leg. He handed me a pair of crutches, and I left the hospital.

When I got home I was famished. It had been difficult enough crawling in and out of the taxicab. Now I had to balance on one leg while wielding pots and pans. In desperation I called a woman I had taken on a couple of casual dates. When she heard what had happened, she came right over.

I had originally met Mary, as it happened, because of a comment I made on the dance floor about my interest in the Amish. Over the din of the amplified music she had cried, "I've always wanted to live on a farm!" She had an impish twinkle in her eye and a shimmer in her movements that I found irresistible. She had the physique of an elf. She was five feet four inches tall and weighed 105 pounds—a weight, I found out later, that hadn't varied in ten years. But she never dieted. In fact, for dinner she prepared three pork chops, and when I passed up the third, she happily ate it herself. What a rare combination of elfin grace and amazonian metabolism.

But what most attracted me was her sparkle. It wasn't just in her eyes—her whole slender being seemed to shimmer and twinkle, beckoning to me like a waving banner.

I never planned it like this, nor would it have happened if I had. For someone who never had found it easy meeting eligible women, the accident provided a remarkable opportunity to get to know one in the most natural way, without any effort or contrivance. For Mary's part, the arrangement was also less nuisance than it might have seemed; she told me it was hardly more difficult cooking two plates of food rather than one, or doing two loads of laundry instead of one (. . . one thing led to another). A beautiful human relationship had flowered despite all the technology surrounding us—or, rather, precisely because of mechanical breakdown.

Mary hadn't been kidding, either, when she said she wanted to live on a farm. But her idea of one was a three-acre lot in Ayer, the second-to-last town on the Boston westerly commuter line. (The train did closely skirt Walden Pond.) She had been house-hunting there in hopes of putting as much distance as she could between her home and her job—a job she disliked but needed in order to pay for the house. As a number cruncher by day, she wore a mouthpiece by night to keep from grinding her teeth. Most of her work consisted in transferring the contents of ledger sheets to ever-changing computer programs, by which she managed larger and larger accounting client pools.

It took a while for my proposition to sink in: a nonelectric hiatus, a telephoneless exile, a life that could pay its own way . . .

"But how do you wash the clothes?" she inquired.

When she asked this, she had been washing my clothes using an electric washing machine for several weeks already. My recovery was coming very slowly . . .

One day I reached in my mail slot and, under the pile of bills and advertisements, found the letter I had been waiting for. It was all set up. In careful, almost childish handwriting, mention was made of accommodations large enough for two.

Her reservations about clothes-washing notwithstanding, Mary was on the spot. Our relationship had heated up well beyond the level of friendship. Fortunately for me, she had never been emotionally attached to her job, and the potted plant in her office simply did not satisfy her hankering for the country. On the other hand, she had a list of unanswered questions: How heavy was the work? How long were the hours? What about refrigeration? What about food preparation? When I pointed out that people have been living without modern gadgetry for thousands of years, she finally gave in, brimming with curiosity to see how they did it.

We tied the knot at St. Paul's Church ten days before the scheduled arrival at our new home. To live in close quarters with a group like this, you had to be properly married. It would be premature to say, however, whether ours was a marriage of convenience.

We decided to shoot for an expedition of eighteen months—

enough time to experience a full change of season. Mary agreed to go along on one condition: that she would get the deciding vote in the decision of where to live after we finished our "fieldwork."

And so, svelte assistant at my side, I set out in the general direction of a still-mysterious clique of manual laborers, imbued with one lone hope: that they might lend me a hand in my experiment. How hard and time-consuming was this life "without laborsaving machines"? And was it one Mary and I would consider leading ourselves? I dearly hoped the exercise would not amount to a sheer test of endurance. What I really wanted to discover was a balance between too much machinery and too little, or better yet, how to arrive at it wherever one found oneself. This knowledge was what modern society lacked and what I hoped my yet-unknown neighbors would provide some clue to.

paring back

After honeymooning in Maine for ten days, we swung west in Mary's red Escort, moving slowly and savoring the sights along our itinerary. But to reveal what they were or how far we traveled would give clues about the destination, which I have agreed not to disclose.

Amish groups relish public scrutiny little, but the settlement I approached was unusually well cloaked. Even though I knew where it was, I still did not know *what* it was, or whether it was really even Amish. One university researcher I consulted, a professional linguist who had been raised Amish and kept abreast of Amish doings, knew nothing of the group. Another insider, Omer Stahl, director of the Mennonite Information Center of Lancaster County, cautioned me not to raise my hopes. Many would-be writers, he warned, had sought to live with and learn from Amish(-like) people and failed. "Be sure to do what you're told," he said. "Don't ask too many questions. And be prepared to get up at four o'clock in the morning to milk the cows."

If it was one thing to theorize about doing with less technology in a classroom, it was another to ponder doing with less in real life. As we drove, the discomfort of uncertainty grew in my gut. The very lack of information seemed to confirm darker intimations. A picture began to form in my head; I tried to suppress it, but it kept swimming in the back of my mind. It was almost too ghastly to articulate: bent-over figures laboring in the muck, adhering to the customs of their ancestors, warding off modernity with hexes and chanting.

Shaking the picture from my head, I reminded myself, "They allowed you to come, allowed an outsider to enter their inner circle. They cannot be so backward." Then I thought of the real reason they had taken me in and my throat went dry: the rowing team. In my letter, as an ingratiating gesture, I had described the endless weight circuits and painful back bending of my chosen sport. No wonder they had agreed. They thought I was one of them.

I braced myself for the inspection. It had been two years since I had last rowed. My center of gravity had shifted from my chest to my waist. Would I be repulsed as a fraud? Or would they simply allow my corpulence to serve as its own punishment, bringing me my just deserts? How would I be brought back to civilization? In an ambulance? In a wagon filled with dung?

Beside me sat my new wife, innocent of what was to come. Where was I leading her? What would become of us?

"Isn't it pretty," she said.

I rustled from my daydreams. The terrain around us was green, misty, and rolling. Farmhouses peeked through the trees, permitting glimpses of steeply angled roofs, fat porches, and muted colors. Homesteads were arranged in a pleasing contrapuntal series alongside the creek that rippled and twisted its way among them, flanked by small rich strips of bottomland. I could see faint green lines in the dirt, shoots of corn or sorghum starting to grow. From what I had read, the land indeed seemed suitable to Amish settlement: hills could not easily be farmed using large-scale mechanization, and the ample woods, besides providing privacy, were a source of heating fuel and wild game.

Something bothered me about the scene, but I couldn't put my finger on it at first. Then I realized what it was: There were no automobiles or pickup trucks in the driveways. There were no power lines connected to the houses. There weren't even any people . . . It had rained earlier; the road was still damp in places. The inhabitants must have been indoors, out of the weather.

The quiet, the mist, the emptiness, the absence of signs of modern life, began to fill me with a sense of desolation. It was as if I were viewing the Earth after some great holocaust had pruned back the population and left the survivors bereft of any modern technology whatsoever. Was I actually getting what I asked for?

Mary and I topped a crest. The curling road bobbed down into a wooded hollow, rising again to a small tumbledown shack on the left. Around the next bend, we came to a large gray home with a steep pitched roof and shutterless windows, set back behind an enormous vegetable garden. For all the vegetation, the lack of a human presence was disconcerting. After curving and bobbing past many similar unpeopled dwellings, we finally found the mailbox number we were looking for. Our hearts quickened.

The house sat about fifty feet from the road; it was a white, sharply gabled bungalow. Though modest of size and design, it had a wide front porch and was set off by colorful flower beds and a broad, lush lawn. Sizable trees provided a canopy of shade, and tuneful songbirds concealed in the bower fluted melodiously. With shafts of sunlight breaking through the clouds, the whole scene shimmered and twinkled as if it were enchanted. Out of the gloom had arisen a fairytale cottage.

These were the living quarters on which our whole experiment depended, the chance on which it hinged. Mr. Miller, the owner, who lived nearby on land presumably adjacent to and in back of this piece, had heard about us from the local church bishop, who in turn had taken over the role of go-between from the man on the bus. (Soon after passing on my name, this helpful messenger had moved to another community in search of a spouse.) Miller had recently bought the property in order to hold it until one of his grown children was ready to move in. In the meantime, he needed to rent it out. The right place was available at the right time.

The circumstance was a piece of astonishing good fortune and made our entire stay here possible. It also eased another potential difficulty. As Catholics we belonged to a religion theologically divided from the Amish and other Anabaptist groups. Long ago the priest Menno Simons had left the Church of Rome in search of a simpler, more personal spirituality. His followers, the "Mennonites," gave rise to another even more minimalist offshoot, led by the stern Jacob Ammann, dubbed the "Amish." Among Menno's grievances with Catholicism was its practice of infant baptism, which he believed engendered lukewarm religious conformity. As a remedy he baptized adults only after they had made conscious assent to Biblical precepts, even if they had been baptized before (hence the word "Anabaptist," or "Re-Baptizer"). Mary and I had both been baptized as infants, once and for all (albeit I as a Lutheran before later switching denominations). It was probably fair to say that on most other matters of Christian form or liturgy, Menno's spiritual heirs still defined their positions as antitheses to Catholic ones.

Given all this potential friction, our landlord's short-term financial interest may have oiled things considerably. At the same time, our theological distance granted us a certain objectivity. Any success that might arise from this venture would be the result of the change in our material conditions and technological approaches, not a prior religious affinity.

In the garden behind our cottage, we spied what appeared to be our own green field, with long faint rows of what I suspected were corn, potatoes, beans, and tomato plants, starting to sprout. I had offered to pay Mr. Miller to have one of his children plant a few vegetables for us before we arrived (it was now June 4), but I hadn't expected a full garden.

The front door of the house was open, and we entered. The interior was almost as refreshing as the exterior. From the moment we walked in, I could see Mary was already enjoying her new home. She looked radiant. The three downstairs rooms were spacious and airy and freshly painted. Front and back porches made them seem even more expansive. And the rooms were furnished. I had been told the house would be empty, so we had expected to spend our first days scouring junk stores and yard sales. But all was supplied. This was a true favor. Catholic though we were, our landlord had gone out of his way to make our arrival comfortable.

As I moved out the back door and around the corner, I almost fell over him. At first I thought it was an elf. He was slender, a bit short, somewhat aged. He wore a wide-brimmed straw hat, blue denim broad-fall pants, and suspenders. His beard flowed thickly in gray and charcoal streaks from his face, and he seemed to stoop from its weight. From under his straw hat, he gazed up at me keenly.

His wife, following behind, half again his size, wore a long blue apron over her ankle-length blue dress and a round black traveling bonnet over her hair. There was a hint of mirth in her eyes, a hint of a quiver in her plumpness. When I looked closer at her clothing, I saw, faintly outlined on it, a rectilinear patchwork, squares of cloth so various and artfully stitched together that it was hard to tell what was original and what was added. These were quilts she was wearing.

Mrs. Miller was the first to open her mouth. "We came over to finish cleaning the house," she said rapidly. "But now that you're here I guess you can do it yourself." With that she gave a little chortle. The Pennsylvania Dutch woman clipped her English and sputtered her sentences like the man at the Greyhound stop.

"Ohhh, it looks fine."

Mr. Miller was peering at me warmly. "Have a nice trip?" he inquired. He talked slower than his wife and seemed genuinely interested in our answer.

"Hit a little rain coming in."

"Yeah, it was raining here too."

Though small and lithe, he bore himself with a certain gravity. After hearing a few more of his carefully chosen words, I no longer thought of him as elfin. The talk was small, but each word had weight; there seemed to be a profundity of interest concentrated in his steady gaze, a wealth of deliberation in his slow speech.

Smiles now and then seemed about to form in the corners of their mouths, and then the lips smoothed out again. I supposed since we were strangers—and Catholics no less—they could not show approval too openly. Or was there something else? Did they see humor in the situation? Did they know something we didn't? Merely everything, of course. That was why we were here.

Before long, the Millers turned to leave. But as he stepped aboard the buggy (which they had pulled up quietly to a backyard bush while we had toured the house), Mr. Miller turned back and admonished me: "You'll be expected to mow the lawn."

The command fell like a brick. My stomach did a somersault. Had the arduous physical regimen already begun? I sighed and chuckled to myself. Too little to go on yet, of course. Perhaps I needed to wait at least to see what sort of lawn mower I would be using.

I remembered that I had not yet given Mr. Miller the rent check, and strode over to him with it—$150 for the first month. I handed him $50 more for starting the garden. He tried to decline it, but after an awkward moment's insistence, he took the money. I hoped I hadn't insulted him. The buggy spun about and they were gone.

We had no formal relationship with the Millers except as tenants. Still, we clung to the hope that living next door would provide informal occasions to learn from them, to put us into the thick of their customs and habits. From my brief communication so far, I knew only the scantest about what those customs were: no electricity, no telephones, no motors, no motor vehicles. We of course expected to abide at least by the same restrictions. It was time to take a technological inventory:

The recently purchased house was still equipped with electricity because Mr. Miller had intended to rent to outsiders. But he hadn't counted on us. We would have the current switched off. As we went through our belongings, we uncovered two kerosene lamps—gifts from our wedding.

Since the water supply depended on electricity, we also needed a substitute for the kitchen faucet. Outside the house, an old pitcher pump sat on a square wooden stand above a cistern, which received the water like a sink. Mary was soon cranking the handle and testing the device's capabilities.

The cistern was good for dish-washing and bathing water. For drinking water we looked to the spring that lay back over the hill. The Millers offered to bring us a container of spring water every two days when they went to fetch their own.

In all of these adaptations, we were conforming to the consequences of turning off the electricity. This also meant there would be no electric stove, much less a microwave oven. Mrs. Miller had hinted that she had a kerosene range to spare, which we could place on the screened back porch for good ventilation.

I lifted my bicycle from the hatchback of the Escort. We hoped to buy another one in the nearest town, so we could travel together. We didn't feel ready to get rid of the car just yet and expected to garage it in the back shed in case it was ever needed, until and if we were able to purchase a horse and buggy.

Mr. Miller stopped by later to drop off the mower. It was a cutting cylinder, two wheels and a handle. I remembered how my grandfather panted as he pushed one like this at his home in Hutchinson, Minnesota, when I was a child. The moment had arrived. I felt a few nervous flutters. The grass was green and tall and still wet in places. Here, it seemed, was the task that would set our whole fate turning: *push* (wheeze)! *push* (wheeze)! *push* (wheeze)! . . .

The whole yard took about an hour and a half to mow. The grass could have been a little shorter and less damp. I had to do some of the thicker rows twice, and to overlap widely on each pass. Pushing took effort in the chest muscles, lower back, and legs. My shirt was soaked with sweat. And yet I was not exhausted. Nor can I say the task was unpleasant. There was something delicious in the feeling of hand mowing through thick turf. As the spiraling blades transmitted my own power into cutting—*snick, snick*—the sense of sinking my own teeth into that chore was palpable, as if to say I had bit off something I could chew. Later I learned that lawn mowing was considered a woman's chore here, and not a very heavy one at that; nine-year-old girls often did it. But for now I felt a small, smug sense of satisfaction.

The next morning Mr. Miller appeared at our doorstep and asked if I could help him carry something. On the back of his wagon sat an old white appliance, a circular washing machine with a crank handle in the center. We brought it into the house, and Mr. Miller left us alone. Mary gazed at the thing in astonishment. I helped her heat the water on the kerosene stove Mrs. Miller had lent us, and she gathered the other ingredients. In went the water. In went the clothes. In went the biodegradable detergent. Mary took hold of the crank and began to turn.

We have since learned that among hand-operated washing machines, the ones with back-and-forth handles work best. A swinging motion, as technological historian Lynn White Jr. has noted, is far more amenable to the physiologies of earthly beings than is a circular motion, which can be seen in nature only in the stars. That is why we later purchased, from a local Amish craftsman, a custom-designed swing-handled machine.

Still, for now the crank-handled machine offered Mary surprisingly little resistance. Having received the tip that two hundred strokes completes a load, she stopped turning after about three minutes. The load was clean. She pulled each sopping item through the rubber ringer and dropped it in a basket.

What a novelty it was. Actually washing clothes—not just pushing a button and hearing a hum, but washing your own clothes without the mediation of distant power suppliers disgorging pollutants into the biosphere. It was really rather cathartic for both of us. We were cleaner in more ways than one. And whenever Mary did the laundry again, I sensed more than willingness in her cranking movements; I sensed zest. This was productive. Thought and action coincided. It was the upper body workout she never had time for on her lunch break, yielding a meaningful end-product. Mary had come down from her counting tower and added body to soul. I liked the sum I saw. When she was done with that first load, we grinned at each other, each knowing what the other was thinking, and I helped her hang up the clothes.

—

But not too much could be concluded from a little hand mowing and bit of hand-washing. The greater part of the manual labors still lay in store. No grant funded this experiment. If I couldn't get food on the table, there was no financial cushion to fall back on. About $1,500 remained from my first year's fellowship money, not even enough to pay the rent for eighteen months. Mary had savings, but these I deemed sacrosanct; they were held in reserve for the day her turn came to choose our adventure . . . unless she returned to the city single. Now that our technological moorings were severed, we were in the same boat as our neighbors. We had to pull our own weight.

I sat down to go over some self-help books we had bought, used, in

Boston: the National Gardening Association's *Gardening* handbook, Organic Gardening Magazine's *Encyclopedia of Organic Gardening,* and a dog-eared copy of an old manual with a grandfatherly man on the cover pushing a rototiller. I had already studied them in-depth; now I reviewed certain sections and took more notes. When I was satisfied with the plan, I dropped to the floor for my old rowing warm-up. After stretching my hamstrings and quadriceps, I phased into sit-ups and push-ups. Jumping jacks followed; then windmills and toe touches. Mary looked on pensively.

Catching a kiss on the cheek, I took my leave—almost forgetting to grab from our mound of supplies the small sack of seeds I needed. Hopping aboard my bike, I set out over the hill. Somewhere on the other side lay the field Mr. Miller had set aside for our cash crop: our best hope for financial solvency.

The Miller place was a little farther from our house than it first appeared—not at the top of the first hill above our cottage, but on another hill beyond that. The gravel road rose for half a mile, crossed a plateau with a little dip in it, and rose up again. As I pumped along, it took a few moments before it dawned on me: I had been taking nearly the same bicycle journey each day to Mary's apartment. The route from my place to hers went up a similarly steep hill, crossed a plateau, dipped a little, then swooped back up again. Central Street in Somerville, of course, had swarmed with moving automobiles, and it was lined with shabby frame dwellings. I also recall a nondescript nursing home, a red-brick apartment building, and an old factory. Here, in place of the houses, was a sloping sward of orchard grass; in place of the nursing home, a row of tall pines; instead of the apartment building, a stand of oaks and maples; and in place of the factory, the Miller compound, a collection of gray farm buildings. Everything had changed but the lay of the land.

It was as if I had entered a parallel world—which in a sense I had. It was an alternate space nearly identical in its geography but somewhat different in its operating principles. What those principles were, exactly, had yet to be revealed.

I began to make out the gables of a sprawling house. Bright pink and orange flowers on the trellises and in the beds alongside the dwelling set off its gray wings and walls nicely. A couple of girls on the front porch were busy at some handiwork, and as I approached, I waved. They waved back.

The barn, an ancient unpainted, sagging structure, towered behind the house roof. As I rounded the bend into the barnyard, I could see chickens rooting in a pen partly formed by the walls of the two structures. To the other side of the pen was a huge low-slung metal shed. Barbed wire twisted and meshed its way among the buildings, and you could see how the inhabitants would move from one space to the next, as through a series of rooms.

I learned from one of the children that Mr. Miller was in the large metal shed. Parking my bike, I entered the building and, after walking around several rickety-looking old implements, came to an interior door. Pushing it open, I saw him crouching over a worktable illuminated by a single shaft of light streaming in from a small window.

I needed a favor. My plan was to sow the half-acre he had allotted me in pumpkins, and the books had suggested burying a shovelful of manure under each pumpkin hill. I wanted to see if he could sell me some manure.

His lips spread to a near-smile, then narrowed again. I braced myself.

After a lengthy silence, he at last spoke. "Do you . . ." He stopped. He appeared to be suppressing laughter. After regaining his composure, he resumed. "Do you really want to do that? Seems like so much . . . *work*."

I did a double take.

"Well," I nervously replied, "would there be a better way?"

There was a pregnant pause.

"We're lazy people, I guess," he said with a sigh. He went on to explain that burying manure under 120 pumpkin hills was a back-breaking proposition. It also might be pointless. This late in the season, the manure might harden in the heat. It could actually stunt the growth of the plants. Better, he thought, to fork it on the surface in rings around the seeds as mulch. Then if rain came, some of the nutrients in the manure would still wash into the roots but not overwhelm them. Mulch of course would also choke out weeds and reduce hoeing. He recommended that I plant the seeds first to get them going, then mulch at my leisure.

As I took in the words, a strange feeling came over me. I soon recognized it as disappointment. I had been gearing up to be the Hercules of our homely days of yore, and Mr. Miller had just stolen my bravado. I felt cheated.

"But yes. You can have some manure."

I stood open-mouthed. Finally I asked, "Can I pay you anything?"

He shook his head. "Come by any time." He nodded and smiled cordially.

Given his tip, planting pumpkins became an easy matter indeed. The Millers had already plowed the half-acre field and sown a crop of corn that had never germinated. All that remained for me was to bike a quarter mile to the site, mark out a ten-foot grid with some rocks, and insert the seeds. Partway into the task, I felt a pair of eyes boring into my back. Mr. Miller's seventeen-year-old son, Ellis, who had pointed the way to the field, was loitering nearby and watching.

"Am I doing anything wrong?" I asked.

"Well . . . this is just me. You may want to do things your own way. But if it was me, I'd be doin' it a little differently." He walked over and pointed to the bag of seeds I held in my palm. "Can I see that?" I handed it to him. "See this?" He pointed to the seed. "It's sharp at one end. Ya' wanna point that end down. The roots come out here." Strange that no manual mentioned this. I did what he told me.

A day or so later, Mr. Miller's two youngest sons, Amos and Caleb, helped me load up a wagon with manure from the barn. In about an hour we'd spread dollops of the stuff around every one of my pumpkin rings. And that was that. The thought was beginning to cross my mind that I might need to adjust my expectations about how difficult this experiment was going to be.

What I was beginning to learn was that life without modern technology need not consist of brawn alone; a deft use of human wits can often do as well as, or better than, scads of machinery or muscle.

Since my first stint mowing, I learned never to wait more than five days in the spring to cut grass, and never to attempt it when wet. I got the hang of fine-tuning the machine, using the little screws the manufacturer had placed on the cylinder just for that purpose. Mr. Miller's second-oldest son showed me how. When you set the screws properly, the blades slice through the turf like butter.

Little tricks like these remove much of the onus from manual labor and add to the sense of physical effort a much finer satisfaction: the magisterial feeling that comes with wielding means precisely fitted to ends. Here, perhaps, is the first of all lessons in the use of power, whether technological or physiological: trimming back the means until only the essential remain; weeding out obstructions, man-made or not, to our goals.

—

Mr. Miller's little tips, hints, and helpful implements continued to flow in our direction.

A day or two later he made another call, offering to sell us a foot treadle for Mary's electric sewing machine. He converted the device right before our eyes. She got the hang of it in no time, and she found the beat of the treadle relaxing. The pumping motion kept up blood flow to the brain, and she felt more alert.

The two youngest Miller boys cultivated the garden with the one-horse cultivator. This spiderlike implement with several segmented legs clawed weeds loose from between the rows, then buried them. When we came over to watch, the boys looked at us and winked. Drawn by a live horse, the contraption performed the most onerous of the garden tasks in a few minutes while providing the user, I found out later in my own attempt, a leisurely walk with a horse.

Two older Miller sons, Ellis and Jed, worked over the pumpkin ground using the disk, an array of sharp-edged steel wheels that sliced the soil at an angle and overturned it. The boys curved slowly by as they stood on the implement and acknowledged my presence with a nod.

Mindful of my need for cash income, Mr. Miller asked his son-in-law, Sylvan, whether he might reserve an additional half-acre of ground some distance away on which to grow sorghum. Sylvan consented. Mr. Miller wanted only thirty dollars for the year to rent the pumpkin patch, and asked only another thirty dollars for the assistance with the horses, the implements, and the boys. Sylvan merely wanted help with his own sorghum crop.

Mrs. Miller, meanwhile, invited Mary over for a day to teach her canning techniques, and they were joined by another nonnative woman, Mrs. Jones. She, by a remarkable coincidence, was also a city dweller who had learned by chance of this unusual group and with her husband and children had moved to the vicinity to learn from them. They lived only half a mile away.

Mrs. Miller, besides lending us her spare stove, offered a huge outdoor kettle to provide a water bath for canned fruits.

The assistance took us off guard. Often it came before we even knew we needed it: a reminder to plant or pick, the loan of a wheelhoe or an axe. We tried to thank our advisers, but they were quick to deflect praise. "Why"—an imperceptible hoot—"no trouble at all." "We were"—slight tone of incredulity—"going by anyway!" "We're just"—arms akimbo—"doing what everyone else used to." And in a final effort to put us at our ease: "Many hands make work light!"

And many hands there were: two each for Amos, age eleven; Caleb, twelve; Judy, fifteen; Ellis, sixteen; Irma, seventeen; Jed, eighteen; and Harold, twenty-two, all swarming about us and related by way of common parenthood. These constituted our immediate neighbors: the Miller family. But they might well have been members of a special Amish ambulance crew assigned to monitor our vital signs and minister to our slightest fluctuations in well-being. They performed this duty with the quiet expertise of trained medics.

At the same time, there was something babylike in their rosy faces utterly foreign to my modern experience, as if they were shepherds just returned from the manger—those slightly dimpled, perfectly oval faces with open, peaceable expressions. The boys were so well mannered that they seemed feminine in their delicacy and sensitivity to our wants.

Through it all Mr. Miller, their supervisor, gave orders while hardly speaking a word. Where he, like the others, was articulate I was in the motions of his hands and eyes. From what little I could tell of the brood so far, I had already begun to imagine that they'd whittled the secret of felicity down to a modest repertoire of deft movements, a brief lexicon of telling glances.

The push buttons of modern technology never seemed clunkier. If this was indeed a parallel world I'd entered, it was not some frozen time capsule. Its residents were alive and adaptable. And in the short time I'd been here, I was beginning to see evidence that a world without modern technology need not be any harder. It might well be easier. And more fun.

lightening up

Something in the Millers' half-smiles suggested that if I was onto some of their secrets, I hadn't uncovered them all. And if my hunches were correct, there should be more to technological mastery than reducing the workload. Something else should expand—the realm of leisure, liberty, or higher human fulfillment. The body, liberated, should find solace in its partner, the mind.

Yet except when their lips curled or their eyes winked, our neighbors remained nearly deadpan. They could have sprung right out of the Amish picture books. The men's beards were scraggly, their brows furrowed. They said little. Even the women maintained an emotional distance somewhere between cautious and grave. The paucity of words gave conversations a dangling feel. After each supposed nonfavor, we were convinced the supply must now be used up; there seemed no outward sign that it would continue. And briefly we would pine for company. Then when the next round of help inevitably came, we took it again as a surprise, and gratitude welled up in us.

Yet although they were not effusive, the Millers certainly never appeared put-upon, much less harried or overloaded, as they tooled by with a new gadget or overturned the soil or fetched our water from the spring or pointed out a telltale cloud formation or mulched their garden. And when we passed their house, at least three of the girls, often joined by Mrs. Miller, would be crouching like toads in the beet patch or bean rows. (It was in that position that I first caught one of them smiling and chatting.) From a distance the casual onlooker must have wondered how anyone could accept such a lot, apparently up to the neck in sheer subsistence activities, uncomplaining, even contented, as if it were nothing. While the whole nation around them was going to incredible extremes to avoid such work, they had gone to conscious lengths to preserve it.

Why? There was this phrase they kept repeating: "Many hands make work light." The statement was true, though hard to explain. Gradually, as you applied yourself to your task, the threads of friendship and conversation would grow and connect you to laborers around you. Then everything suddenly became inverted. You'd forget you were working and get caught up in the camaraderie, the sense of lightened effort. This surely must rank among the greatest of labor-saving secrets. Work folded into fun and disappeared. Friendship, conversation, exercise, fresh air, all melded together into a single act of mutual self-forgetting. But why didn't the Millers act friendlier or more outgoing in the first place?

We barely had started hoeing our own vegetables when Mrs. Jones, the Amish apprentice from the farm on our other side, her husband, Nate, and four of their children descended on us. They, like the Millers, were all business at first. They sternly warned that winter would soon be here and that we needed a larger garden if we were to get through it. Brandishing hoes and packages of seed, they helped us add two more rows of sweet potatoes, two rows of black-eyed peas, six extra rows of corn, and several rows of beets. Mabel, the Joneses' only daughter, seeing Mary pull out weeds by hand, gave her a free lesson in hoeing. Mabel was only eight, but quite nimble. Mary earnestly took heed. (The trick was to take shorter defter strokes, rather than golfer's swings. With her small frame, Mary was in need of this pointer since a particularly hard chop to the ground could have a pole-vaulting effect.)

And then things turned around in that inexplicable way, and we were all laughing and talking and having the liveliest time. The Joneses appeared to have learned a thing or two from the Millers.

When the Miller boys swung by with their horse to cultivate the whole garden again, they appeared placid as usual, but our sappy looks of gratitude did raise from Caleb a self-conscious half-smile.

We decided to visit the Joneses to see if they needed help in return, secretly hoping to kindle the friendship. Their place was a rustic two-story dwelling Nate had built from a design he had found in a book, using some manpower he had managed to barter from the Millers. It looked like the house of Ma and Pa Kettle. Nate said it cost him only $1,500; he had made it mainly from used materials.

The Joneses were canning beans when we arrived, so there was little we could do but watch—not enough room to assist. But we could also chat, and we did discover that their reasons for living this way were purely doctrinal; that the Bible, for instance, required all women to wear head coverings. Nate, it turned out, was fond of quoting passages from Scripture that seemed to support practices like this. Another example he gave was the commandment to observe the Sabbath, which he upheld by worshipping on Saturday and working on Sunday. But after gingerly disentangling ourselves from these issues, soon we found ourselves talking and laughing again. The social magic could overcome something more sober than labor—dogmatism. Maybe we had discovered another side to the coin of free-given assistance: Since the purpose of getting together was not social, there was no pressure to "like" or "be liked." Just as conviviality had taken up the onus of the work, the work took up the onus of the conviviality.

This indirect approach to companionship got us out of our shells and inexorably entwined our fates with those around us. Despite dissimilarities in social background and education, we found ourselves back at the Joneses' again and again, enjoying each other's company. The habit soon became a tradition. Manual labor was both the occasion of the parties and the substance that got us mixing and conversing. It was the nonalcoholic cocktail. I was so intoxicated socializing with the Joneses, alas, that I found it difficult at first to make the effort to remember and write down afterwards what actually was said. I was having so much fun that I forgot that one of my purposes in coming here was to do research.

Physical work, then, served more than one function. Besides putting bread on the table and vigor into the physique, it also provided a special social elixir. And this, at the same time, explained why you couldn't appear too eager to see your neighbor. You had to keep up the pretext of labor or the whole arrangement would collapse. Friendship was something you could only sidle up to obliquely. Or maybe it would be better to say that you let it sidle up to *you*.

Given this principle, the problem of taking notes and recording thoughts soon became a serious one. Being of course a form of conscious capture, it was contrary to the whole undertaking. The difficulty extended even to such basic facts of daily life as the work itself. Yes, the labor . . . We were supposed to focus on this, but it was often easy to forget we were even doing it.

Well, there was hoeing, and there was harvesting, and there was canning, and there were field crops. There was just a little bit of everything, but it was getting interrupted all the time, and the demands were unpredictable. I tried to maintain a daily log of events, but naturally some items were missing. And often enough, just as I was sitting down to make a note, there'd be a knock at the door.

Mrs. Miller and her oldest daughter interrupted my train of thought one day, bringing us some extra peaches.

"You brought these for us? Thank you so much!" I cried.

"We were just going right by anyway!" Mrs. Miller responded. Peaches luckily were high-acid fruits, so we didn't need to pressure-can them. We could dispose of them quickly in the large black kettle Mrs. Miller had loaned us, which sat in the backyard atop an old oil drum with a door cut in it. The drum served as a firebox. We would boil the peaches in quart jars, right in the kettle, for a few minutes and they would be done.

First, however, we had to prepare them—to wash, peel, trim, section, and pit them. The greater part of canning was like preparing a never-ending salad or fruit cup. Fortunately peaches were not nearly as difficult as beans, with all their little black spots, tips, and stems that had to be cut out. Corn was also rather involved, with the husking, silk removal, excision of wormy regions, and knifing of kernels. But far and away the greatest time-monger of the vegetables was the pea—the sweet, innocent, small, round pea. Every pea came from a pod, and every pod had to be cracked, and by the time you cracked a large bucketful of peapods, what remained was nothing by comparison—

about enough for a mixing bowl. So it took mountains of peapods to get a reasonable hill of actual peas, each obtained by the muscular cracking of a leathery, sometimes stubborn green shell. Somehow the time-to-output ratio seemed wrong. Secretly I began to wonder whether peas were worth it. Was this the one job that was too much work? The Amish woman who let us pick peas from her garden after she had planted about three times more than she needed said simply, "I like them." But perhaps she also might have missed the annual social ritual that shelling peas summoned forth. And I could not discount her own statement. The fresh raw country pea gave out a green gush, a sweet explosion, like no pea taste I'd encountered. But so it was with the tomatoes and the corn—the taste exploded in your mouth. And we discovered potatoes. I'd always assumed that the potato was an obligatory starch accompaniment. Now a bite of a buttery young potato fresh from the ground was close to a taste of heaven. Vegetables were becoming the main course. For that matter, mealtime itself was assuming a new place of importance in our lives. It wouldn't have taken much thought—if we had wanted to put in the thought—to realize why.

The signs of how hard we actually were working often came surreptitiously like this. Ravenous hunger. Mmmmm. Another clue was the pants that started falling down. These incidences provided fleeting signals from which one could infer hardship. This was the exact opposite of the modern gym, where people keep constant track of physical effects in mirrors and dials that actually focus awareness on the exertion itself.

Anyway, as for the peaches, Mrs. Miller and Ada decided to stay long enough to help dismember them (even though we hardly were deserving of still more service). With their help, the task began to assume the same festive air as a quilting bee—or what I imagined one would be. It took extra effort on Mrs. Miller's part not to smile and fraternize, and by the time she departed she had lost some of her stoicism in this regard.

Another time, as I was deep into recording some insight or another, the persistant rapping sound came again. Caleb and Amos, the two youngest Miller boys, were at the door, stringing a cow behind them.

"Hello, Mr. Brende," said Caleb. "This cow's just calved and Dad, well, he thought you might like to have 'er 'round, sort of to learn how to milk 'er. Have you ever milked a cow before?"

Inwardly I groaned, *Not now*. But to Caleb I answered that I had once done so on a school field trip, and I followed him toward the barn.

"She's only been milked a couple of times," he explained, dragging the animal for all he was worth. The cow was stretching away from him and craning her neck as he pulled. He was but a rail of a boy, and it was a marvel the cow didn't pull him over. "She's a little skittish. Whups!" The rope slipped from his hands just as we crossed the threshold of the stable. The cow made a circle inside the interior, and stormed us for the exit. "Quick!" Amos darted sideways to swing shut a lower door-piece but seeing how small he was, I dashed over and planted my body next to his. The cow drew back, and Amos secured the door. "She's missing her calf," Caleb said.

Now to grab the cow's leash. I lunged; she dodged. I stalked; she backed off. The game continued. When Caleb and I finally went at her from two directions at once, she ran up the middle with a tumultuous prancing of hooves. This cow clearly had missed a career in the NFL. Despite my initial reluctance to follow Caleb to the barn, my heart was now pounding, my eyes glinting. I was caught up in the game. Finally, with one last desperate dive and scoop, I caught hold of the rope. Caleb tied her up.

"Now," Caleb said, seating himself on an overturned plastic bucket, "see these two dugs here? I think she's dry on this side." He was tugging on a side of the udder that looked withered. "But you can still get a lot out over here. See?" He began pumping rhythmically from a plumper area and milk squirted into the bucket. In no time two inches had accumulated. The cow stepped forward, and Caleb automatically shifted the bucket to avoid the foot. "Now you try."

Having come this far, I was a bit bolder than I had been at first, and being reminded of how to clasp the teat between thumb knuckle and index finger, actually saw my first sizable squirts of cow milk. A certain reckless ease was the secret. Thinking too much about it, worrying that the cow would kick, reduced the stream to driblets. "Does she kick?" My stream started dribbling.

"Well, when Helen milked her yesterday, she sent her flying back five feet."

My dribbles became drops. "Really? Is she all right?"

"Yes, just got a bruise. As long's you stay close in right here, you'll be okay. Don't leave the bucket next to her or she'll step in it." He next spread a piece of cloth over the bucket to prevent flies and dirt clots from getting in.

Milking proved the point once again: that the work, while necessary in itself, was an occasion for something else. In fact, focusing too hard on *milking* was fatal to it. If you became fastidious and thought about the mechanics of the job, the cow would sense you tensing and dry up.

The daily task of rounding up the cow really served as a chance to practice football maneuvers, whistle, or pal up with a next-door neighbor.

The next time the knock came, Amos, the youngest of the Miller clan, was alone at the door, or so it seemed. He began to stammer. "You . . . might . . . want to check your garden right away." Then from behind his back he pulled out a large green bean. "The Blue Lakes are ready. Uh . . . uh . . . you probably don't want to let them get this big. They start to get tough." Amos handed me the bean.

"Thank you."

I followed Amos to the garden. Beans were not the only things getting out of hand. So were the redroots, lamb's-quarters, and fugitive orchard grass. Admittedly there were a few farm chores that took real concentration, such as bean picking and weeding. (Note that cultivating the garden with a horse removed weeds *between* the rows, mostly not *in* the rows. The remaining ones had to be hoed by hand.) We tended to do our other tasks first before moving to these. But there was a problem with this default scheduling: the risk that before you finally got around to them, it would rain. Weeds in particular must be hoed on time, whether you have the will to hoe or not. For two weeks, rainfall had interrupted our weeding while making the weeds grow faster.

We stood by the garden and gaped, trying to get a sense of the scope of our predicament. It appeared we had neglected another truism in the economy of life with less technology: an ounce of prevention is worth a pound of cure.

All of a sudden I looked up and saw Mr. Miller standing across the garden from us. He reached down and pulled up a large weed with a long thick root that resembled a twisted beet. "Are you growing these to eat?" he asked. Before we could answer, he continued, "Do you know what this is?"

There was an uncomfortable pause.

"A redroot?" Mary offered, hesitantly.

"Right, very good. A weed."

It was *possible* there was mirth on Mr. Miller's countenance, but we couldn't tell from this distance. Either way, the point was made.

But where should we begin? There were beans going to waste too. We decided to save a few beans first, then move on to the weeds.

Beans, alas, took almost as much concentration as weeds, and picking them was our second-least-favorite chore. It was clear when a bean was a baby or a giant, but everything in between seemed shiftier. When you picked a big one, then the next seemed too small, and you'd hesitate—was it? Then later, you'd have worked yourself down to picking a small one and the next seemed too big—but was it? (Once they got big and leathery, beans were best left to hang.) Then you'd realize that some of the small ones you'd rejected for being too big were bigger than some of the big ones you'd rejected for being too small. It was all complicated by the fact that they were partly hidden in the foliage and sometimes resisted the yank. In the end they just swam before us.

When this happened, our minds, awash in the sense of futility, drifted from the scene. Conversational nothings began to pass between us. We began to talk about this and that, a funny thing we remembered, a dream one of us had had the night before, something someone had said. Later—I don't know how much later—we would find ourselves at the end of a row, our buckets full. The picked crop was all approximately the right size. Where did they come from?

Bean picking, we discovered, was a sort of obligato, a repetitive figure dull in itself, providing rhythmic accompaniment for the larger symphony of consciousness, with melody of voice floating amid violin breezes and piccolo birds all rising heavenwards under the white-splotched blue dome of an orchestra hall. Beans kept time for two lovers to make beautiful music together.

And so it went with the weeding. Each new weed, staunch and muscular in its defiance, posed a new mental knot to untangle. One was fat and scrappy; the next lanky and sinuous. One's roots might encroach upon the tomato vine; another's stalk might mimic it. And then, without warning, our bodies went at it solo while our minds soared into the skies. How strange was the oscillation. One minute we despaired of ever finishing a row. The next, Mary asked, "What was so bad about Descartes, you were saying?"

When our two-day marathon in the garden was over, we came to a strange realization. These tasks that we had most put off—this bean-picking and this hoeing—were why we had come here.

The inventors and engineers may have overlooked something: physical labor is self-automating. Somehow in the labyrinth of neurons that interlace mind and body, certain tasks can be learned and repeated semiconsciously while other tasks are added. Natural opiates called endorphins must enter in next, numbing awareness of the lower functions, liberating the mind for higher ones. The human body contains its own well-oiled "laborsaving" mechanisms.

But this mechanical metaphor is only that. Music may come closer to what we really experienced, for in the work many of our fondest longings harmonized. Still, what we performed in the garden could not be reproduced in a piano practice-room. For life as we knew it now, Chopin was only a prelude.

Perhaps this, then, was the secret the Millers had kept to themselves and were too busy—and tranquil—to talk about: it was easier to do than to describe.

artfully answering nature

Just when we thought we'd overcome our biggest hurdles, disaster struck. Our little farmstead rested snugly in the lap of a partially wooded hill that, under normal circumstances, afforded mild protection from the weather. On one overcast afternoon, however, the hill and woods simply disappeared. It was as if a curtain had fallen. The torrent soon hitting our metal roof was so heavy it drowned out the sound of our own voices. Over the next three days, the deluge abated only slightly. We found out later that it was the residue of a major hurricane that had blasted one of the coasts.

A foot of a hill is not a good place to find oneself after a rain like this. When it ended, our back porch, chicken coop, woodshed, and farrowing house oozed with the runoff from our earthen protector.

We were truly at a loss. We didn't know where to begin. As we listlessly gazed at the damage, a buggy pulled up quietly in the backyard and Mr. and Mrs. Miller, with two older sons, spilled out. By the end of a busy morning, Mary and I, together with the house and outbuildings, were aired out and good as new. With little fanfare, our rescuers in their two-seater disappeared over the hill.

It was as though some sort of reverse-hurricane had struck: four smiling cyclones, bending and gyrating about the property, airing out rugs and casting off debris while spewing out a bit of cheerful chatter. The phenomenon was swift-moving, decisive, unexplained. As they left, Mrs. Miller uttered the words, "I hope we didn't interrupt your schedule!"

The event brought to mind a theory I'd read about the "moral economy of the peasant." The political scientist James Scott, observing a lively practice of neighborly aid among Southeast Asian villagers, concluded that the absence of modern technology made interdependence a matter of sheer survival. The fickle forces of nature thus triggered a counteracting human solidarity, which itself fed a yearning for togetherness that seemed natural. The process appeared to begin and end with nature.

But I suppose that there was an important difference between the Millers' aid and Scott's peasants'. The peasants had not chosen to go without modern technology. The Millers, at some remove, had. Certain Amish groups in the past went so far as to outlaw lightning rods, for fear of diminishing occasions of spontaneous barn raisings. To this day our neighbors forbade the purchase of insurance policies for similar reasons.

On second glance, then, the situations of the Millers and the peasants were not quite alike. The Southeast Asian peasantry got together in the name of self-interest to better their common lot. Our neighbors had worsened their lot so they could get together. The motivations were opposite. A certain artifice was present in the Millers' seeming spontaneity. Or maybe the word was artfulness.

Early in the summer Mr. Miller, like the Jones family, had warned me about the coming winter and gave permission to gather fallen limbs from his woods. But winter seemed a long way off and I procrastinated. Then one day he appeared, bow saw in hand, and began severing large branches from the tree by our kitchen window. It was a perfectly good red maple. I went out and conversed with him as the boughs came tumbling down. He mentioned something about making light for the kitchen sink and protecting the cistern from growing roots. Soon I found myself working beside him, chatting and cutting up a stack of logs conveniently felled near the woodshed.

As he pulled away in his buggy, I reconsidered what had just happened. Did he really mean it about the dangerous roots? I gazed at the stricken tree—nothing remained but a stick—and with a start I realized what Mr. Miller may really have done: created his own mini-disaster. Namely, he had simulated a lightning bolt. Adversity was not providing the spur I needed to get going, so he went ahead and played nature's part for her. And as we worked I'd felt a nice spurt of camaraderie.

For all the mysteriousness of the Millers, evidence mounted of a deeper mastery, an ability to ride the bumps and turns of circumstance and to convert or redirect them to higher ends—a kind of metaphysical hydroelectrics. Mr. Miller's finesse, then, raises an intriguing possibility for those of us still abjectly dependent on technology: that a more natural way of living can, in some manner, be artificially contrived. We need not wait for chance or disaster.

Even the person I knew as Mr. Miller, I soon learned, was partly an artificial construct. It was Nate who informed me. My jaw fell when I found out. Mr. Miller had come from Lancaster County and had once lived in a setting almost as modern as anyone else's. He didn't belong to the group here but, in absentia, to a community of more up-to-date Mennonites still residing in the Boston–Washington corridor, whose male members ordinarily shaved their faces. His brethren used steel-wheeled tractors, electricity, and in-house telephones. While similar to the Amish theologically, they were more progressive technologically than even the most liberal Amish settlement. My landlord's repertoire of skills was broader than I first imagined.

One day in his parlor I discreetly brought up the subject. He smiled and sighed, as if to acknowledge the fact that at last his secret was out. There was a pause in which he appeared to be carefully choosing his words. "I did it," he said, his lips quivering slightly, "for the sake of the children."

As conditions in his former habitat deteriorated, he explained, underage drinking, teenage pregnancy, and even drug use had steadily increased. In the suburban world of cookie-cutter houses, strip malls, and automobiles, parents were losing control of their offspring. Instead of working together on the farm, family members went different directions—fathers to factories, mothers to baked-good stands at highway outlets, boys to traveling construction crews. Money and wheels flowed freely. Opportunities for fun and escape multiplied. One in three Amish youth from Lancaster County were leaving the fold. The attrition rate was even higher among youth from his Mennonite denomination.

He gazed past me into the distance, as though trying to keep his mind on the lofty ideal to which he aspired while minimizing a kind of embarrassment.

He, after all, had never officially resigned his Pennsylvania church affiliation, a requirement for joining the present enclave. He did not even subscribe to all the tenets of the local theology, but he made the effort to attend services and to participate where he could in other aspects of community life. Without being too obtrusive, he even occasionally discussed points of doctrine with church leaders in hopes to win them to his positions. Though not a member here, he had cannily placed himself near it so as to partake of its benefits, and even to subvert its doctrines.

Miller bowed his head after these disclosures as if in partial atonement for what he had done. But I couldn't very well hold against him a crime so similar to the one I was committing myself.

If Mr. Miller was doing it, if Nate Jones was doing it, and if I was doing it, this raised a whole new question: was anybody else? After spending a little more time in the area and meeting a few more of the neighbors, my head began to spin. The real question was, who wasn't? What I had taken to be a homogenous Amish collective was actually an aggregation of Amish, Mennonites, and mainstream Americans from all corners of the country, bearing a variety of religious viewpoints, joined by one converging aim: to reclaim their lives from machines. At last it came home to me why the man on the Greyhound bus had been so evasive about the group's identity. It wasn't really Amish at all. It was still partly up for grabs. It meant different things to different people. Certain core members maintained doctrinaire Amish positions, but even they gave sermons in both German and English—a rare concession among Amish communities. The real identity of the community was still in formation. Although they didn't correct people who called them Amish, they themselves preferred another name that, however, was also somewhat misleading; and I cannot repeat it without giving them away.

Amish people sometimes referred to motley associations like this as "chowchow" groups, after a pickle relish made from a variety of odd vegetable bits. Let me suggest a more flattering and descriptive substitute. The name "Minimites" comes to mind, in honor of both the history of Mennonite nonconformity and their current predilection to gain a maximum of ends with a minimum of technological means. Like Amish people, the Minimites were selective about technology. But unlike many Amish people, they understood well the point of being selective and were less prone to clutch at old-fashioned practices for form's sake. To them, minimation was a principle of fuller living that anyone could put into practice, not a license to withdraw into a closed circle of the chosen or the inbred. Yet in this college of amateur economists, you couldn't tell who was who by looking at them. They spoke both English and Pennsylvania Dutch in the presence of outsiders. And the former appliance salesmen's beards were no shorter than those of tenth-generation descendants of Alsatian Anabaptists.

The Minimites' general store sat in the center of the community. Mary and I had found it a treasure-house of useful homesteading merchandise, attended by a friendly farmer with seven daughters. And with its location and homey front porch, shaded by a couple of overhanging trees, it made a natural site for informal meetings.

One afternoon I came upon a group of men huddled near the store who to all appearances were Amish. Each wore a straw hat and had a beard.

I paused to overhear what the men were discussing: beards. It seemed as though a report had come to them about an Amish man from Lancaster with a red beard so long it separated into two branches that tapered only at his knees. The news aroused a murmur of disbelief. Finally someone harrumphed, "Beards may rightly be trimmed."

"But God put them there for a reason!" cried another (with a slightly longer beard). "What if you rented a house and the landlord grew a nice big lawn front and back, and you decided to mow just the front third down as low as it would go, with the dirt flying! And you kept doing it day after day! What would your landlord think? Two Mennonite boys came to the door, and one asked me by what right or privilege in the Bible do we let our beards grow.

"I told him, 'It says that God made Adam and Eve, and when He made Adam He made him with a beard the way he is now. If He gave him a beard He must have done so because He wanted it that way. Now by what right do you cut your beard off?'

"Then the fellah who was with him said, 'This man is right.' Boy, you should've seen the look on his friend's face."

A listener with a reddish-brown beard chuckled. "People ask me on what ground I let my beard grow. I say"—he pointed to his cheek—"on this ground."

"There is a Mennonite church where they don't fellowship with anyone who lets their beard grow."

"Do the Beachys still grow beards?"

"The younger ones, no, but the married still must," continued the main speaker. "Some of them have got it down just to a shoelace. That's how it looks. Once when I went up to Hiawatha for a funeral, there were some of them sharing a motel with us, and it took them three quarters of an hour in the bathroom just getting their beards trimmed."

(I happened to know that the "Beachy" Amish were a liberal subsect who drove cars. Beachys viewed Old Order sects like this as hopelessly uninformed about salvation.)

"In George Washington's time no one wore beards," interjected a fourth man, "but by the Civil War they did, with every president until William Henry Harrison. Then they stopped."

"When I was at the neighbor's I saw a newscaster on TV give a speech who wore a nice beard."

Someone pointed out that one Mennonite church would allow bearded Beachys to preach but not unbearded.

"Boys without beards represent a youth problem," concluded the most vociferous member of the group. "When they have beards, there's not much of a generation gap."

The last comment alluded to the fact that this settlement permitted beards among unmarried men—an unheard-of concession in the Old Order.

As the conversation wound down, I wondered who of the speakers were "native" and who transplants. The man who spoke of the presidents seemed suspiciously well educated, but then the fellow who clamored for trimming may have also revealed a certain heterodox leaning.

But whoever they were, one thing seemed clear: some of my first impressions may have been too hasty. Their beards were not mere customary trappings, nor were they features of a costume. The roots, if you will, went deeper. Behind the discussion of facial hair lay the attempt to tease out a common vision of the life they were attempting.

Liberal Amish groups like the Beachys consider such matters cosmetic and trivial. What difference does the number of buttons, color of clothing, or length of beard make to a God peering into someone's soul? But to take the discussion in this way is to misconstrue it. I sensed a deeper theology—perhaps a theo-ecology. One religion may focus on God and the supernatural. Another may make nature a God, like the Druids or Science with a capital "S." The unique contribution of this bunch was to connect the two—to link heaven with nature, covert cause with overt effect. In a world saturated by God's influence, (or substitute here your own word for the hidden impetus underlying and uniting visible phenomena), certainly nothing is unimportant; everything in some way impinges on everything else. Admittedly there are some fuzzy areas, but that was precisely where the discussants came in. In the appointed order of things, there is a middle realm of human influence and refinement. A balance must be struck between wild overgrowth and bald control, a humanizing amount of trimming and shaping.

If I had joined the discussion, I might have put in my own two cents' worth. I knew about Kalahari bushmen and the fact that they couldn't even grow beards. Maybe in the divine economy facial hair served as a protective option for men in colder climates, to be shaved off in the summer. I put my hand on my chin . . .

. . . Then again, maybe not. Without electric lights and replaceable razors, shaving every day is highly problematic. I left the huddle, stroking my new whiskers thoughtfully.

the ram

One warm summer afternoon little Amos looked up at me again from our back stoop. At eleven, Amos was Mr. Miller's youngest son and most frequent errand-bearer. His face was so open and ingenuous, it shone; the mere sight of him brought on a feeling of peacefulness. He had joined his older brothers the day they showed us how to use the one-horse cultivator, with its adjustable prongs that widened and narrowed to fit the width of the row. And he had personally advised us on the picking times not only of beans, but of corn and cabbage and potatoes as well. Despite his age, he possessed the quiet confidence of a seasoned produce manager and, yet beardless, the diminutive authority of his father.

Now he politely asked, "Would you like running water?"

Had I heard right?

I knew there was a spring somewhere in the woods on the other side of the hill, but it took electricity to pump the water to the house. Since we had forgone household current, I didn't understand Amos's point. Now he explained: Electricity was not the only source of power available. There was also the *water*. Since it ran down a little hill, the power it generated could be used to pump *itself*. The idea was ingenious and elegant, and I had never grasped it until now. A certain device called a "ram" had been invented to combine the two uses of water: a water mill for pumping water. By sending the liquid through the small mechanism, a certain percentage would flow back up the hill, over the crest, and down into our house . . . and make our lives simpler.

Though I was intrigued by the technology, the offer made me uneasy. Had Mary and I come this far only to mechanize? Was Mr. Miller trying to tempt us from the course?

But on a moment's reflection, I realized my error. Since when had the Millers abstained from technology? The evidence was everywhere and inescapable: the cultivators, the buggies, the canning equipment, the countless other basic utensils and implements. Evidently technology itself was not taboo, only technologies that interfered with this plain sect's aims. Put positively, our neighbors chose devices they thought would benefit them—the minimum necessary to maximize their ends.

And the amount was still minimal. As these frugal people well knew, technology, and in particular motorized machinery, always brings a cost, whether up-front, in dollars, or long-term, in repairs, fuel, and maintenance. More important, even at low monetary cost, experience showed that such gadgetry can easily interfere with the delicate dynamics of the human welfare it is supposed to promote.

Still, to use minimal technology is not to use none. Nor is the minimal amount some arbitrary quantity. It is the minimum one deems necessary to one's aims, and this may sometimes be more than a little; it may be quite a little.

Rams are very sophisticated mechanisms, yet they were by no means the most elaborate or substantial technologies the Minimites used. I noticed in nearly every local kitchen a big black, shiny cookstove with a little insignia on the front bearing the words "Pioneer Maid." It was an invention of two Amish brothers from Canada, and it was more than an ordinary stove. It was the first-ever application of the principle of airtight combustion to wood-fired cooking. This made it the only notable advance in wood cookstoves in at least one hundred years, probably since the introduction of cast iron. Besides being efficient, the stove was versatile. It could cook, bake, maintain a hot water supply, dry vegetables, and heat 2,000 square feet of living space all at the same time. For the local housewife, it was an all-purpose appliance that met most of her heating needs at the touch of her fingertips.

Affiliates of the local group had done their share of innovating too. The rule against motorized equipment, sometimes thought to be a common feature of Amish life, was actually quite unusual, and it presented the community with certain technical challenges. How, for instance, should they saw lumber? For a while they managed by hiring the services of an outsider who operated a portable band saw, but this made them dependent on expensive machinery that wasn't always available.

To remedy the problem, two brothers, both Minimites with a strict Old Order upbringing, decided to build a water mill. A farm they had recently purchased had some rapids running through it in a hollow spot that was big enough to hold a small lake. The principles of mill construction were a bit beyond them, so they plotted out a rough design on paper and brought it to a mechanical engineer. With his help, they were able to carry out the plan.

I beheld the result: the mill wheel, about eight feet in diameter and welded of sheet metal, spun beneath a small earthen dam that held back a pond spreading over about half an acre. As water trickled over it, it quietly turned and powered a rotary saw housed within the mill building overhead.

On a stroll through the neighborhood, I happened to bump into one of the mill-owner brothers and struck up a conversation. During the course of the discussion, Gideon recalled another more important technical problem he had had to overcome when he first arrived in the area ten years earlier. As he spoke, his eyes became animated and his gestures emphatic.

The former landholders, he cried, would "take a piece of land and farm it in corn five years, five straight years! The ground was hard and lumpy. All the worms and other things that live in the soil—the old worms, they just leave. They used chemicals to control the weeds and brought in all this heavy equipment. The equipment pounds down the soil. You get a six-inch-deep crust—what can you do?"

"The first time we tried a field," added his companion, "ran a disk crosst it, you couldn't even see where it had gone."

"It was just so hard and lumpy," repeated Gideon, shaking his head. "The way we farm it really makes a difference," he went on.

When he and his brother arrived, they introduced techniques their Amish forebears first used hundreds of years ago in Alsace, namely crop rotation and the spreading of manure. In a few years the soil was rich, loamy, and full of living things.

In the midst of a sentence, Gideon paused. He pointed to an intricate web stretched almost invisibly between two cornstalks. In the center of it sat a big yellow spider. Gideon grinned. "This old spider, if you just touch its web, it will come out and defend itself—sit in its nest and shake back and forth." He did an imitation: mouth agape, hands clutching the air, body swaying back and forth. Then he touched the web, and the spider bore him out.

Organic cultivation might be called a way to humanize the soil. But these people were not above extending this craft with technology. In pursuit of a lead someone had given us, Mary and I took the Escort on a half-day drive. When we got to our destination in the next state, we beheld the cleverest instance of minimation we'd seen yet. A local blacksmith had taken the mechanical assembly of an old John Deere motorized baler, the 24-T, and customized it with special gears, drive-chains, and a steel wheel from an International tractor, the W-30. From all these ingredients, he had contrived a horse-powered hay baler. Unlike the balers used by most Amish groups, which are drawn along by horses but have gasoline-powered engines for the actual task of baling, this one was powered entirely by horses. Weighing less, it reduced compaction of the topsoil while the team of work animals added more organic matter.

The blacksmith who had devised the adaptation looked at me warily when I knocked on his door. But when I told him why I was visiting, he smiled broadly. Where were we from? he wanted to know. When I answered, he asked me if I knew of so-and-so Stoltzfus. I did, though vaguely, and with that he insisted we visit that fellow's brother, who lived just down the road from his shop. After chatting for several more minutes, Mary and I sought the second man out.

In the kitchen of a tidy white farmhouse, we found the fellow in the midst of lunch with his daughter's husband and children. Soon, seated at the table, we were nibbling grilled smoked chicken breasts, buttered noodles, coleslaw, pickled beets, mashed potatoes and gravy, and at the end of this banquet, the *pièce de résistance*, peach upside-down cake with real whipped cream heaped on top. Conversation ran the gamut of friendly inanities, touching on even a few (vague) remembrances about the brother we were supposed to be talking about.

At the end of the meal, mother and two grown daughters (one having the day before given birth to a baby whose arrival it turned out they were celebrating) sang us a farewell hymn in three-part harmony. The intonation was perfect.

As we pulled out the driveway in our car, the son-in-law leaned over the window and detained us another thirty minutes with more inquiries about life in our parts. I don't believe I've ever met someone who gave out such sighs of contentment or radiated such beams of satisfaction. We were able to leave only by inching the car forward until we were out in the road.

—

When the quantity of machines shrinks, another area of human realization expands: skill. Strange that our mental picture of life with simpler technology is peopled by drudges and unskilled laborers. That is doubtless a projection of our own experience—the mindless repetition brought about by automation over the last two centuries. Mary and I were discovering now that it wasn't the sheer physical burden of unmechanized labor that was daunting. It was the skill. To make matters harder, skill was not concentrated in a single specialty but scattered in dozens of little knacks and hundreds of bits of knowledge, all foreign to the button-pusher. On top of all this, the foremost skill was balancing and integrating all the little bits into a single livelihood. When I told one young farmer how impressed I was by how much he knew about so many things, he looked at me, grinning, and said, "But I never do one thing long enough to get good at it!" Another time, as I observed some Minimites with an Amish background tearing down an old house to reuse the lumber, one turned to me and asked, "Do you think you could disconnect the electric lights for us?" I suppressed a laugh. I no more knew how to dismantle electrical connections than how to tie a double half-hitch. He had assumed that, as a representative of technological society, I must know how everything in it worked. And why wouldn't he? In a society where machinery has not displaced human skill, people still do. By minimizing technology, our neighbors maximized human know-how.

To get a sense of what went into the vast repertoire of skills they took for granted, consider what it was like for me to learn a single one: operating one of their chosen non-automatic gizmos. Shortly after Mr. Miller had arranged to have me grow sorghum with his son-in-law, I biked over to Sylvan's to see what needed to be done. I found my sorghum collaborator in his barn. He had big bushy eyebrows and looked plump and thoughtful as a hen. When he saw me through the doorway, he arched his brow in mock alarm as if to say, "Here we go."

But he only said, "Ohhh! Well, well. If it isn't Eric Brende. I was just about to work your field over with the cultipacker. Actually, you only get sixteen rows of the field, at the upper end." He was standing next to a pair of horses and proceeded to throw on their harnesses. Leading them into the barnyard, he hitched them to an odd device that consisted of a heavy-looking metal roller about eighteen inches in diameter and eight feet long, with a small platform above it for the driver to stand on.

"What is a cultipacker?"

"It pats down the ground so the seed don't blow away."

"Oh." My voice fell. "So you've already planted the sorghum?"

"Yes."

"Do you think I could still help out some? Could I do the cultipacking for you?" I thought I saw Sylvan's eyes roll slightly. "After all," I quickly added, "I ought to pull some of my own weight here. I have worked with horses before." This was a partial truth: I had driven a single horse pulling an open-air buggy once in Lancaster County.

He thought for a moment. "Well. I really don't see why not, if you want to." He clucked as if there were some joke in his reply that only he understood. I was following him now as he led the horses and contraption down a sloping pasture from his barn. We came to a small opening in some trees, then entered a round field of dark, recently tilled earth completely surrounded by woods. "Just be careful not to fall forwards," he said, handing me the reins. He smiled good-naturedly and ambled back toward the barn.

He was leaving already? He evidently had believed my line about the horses. I looked at the cultipacker and frowned. How heavy was the roller, if I should . . . ?

Hesitantly, I stepped onto the perch. It was about big enough for my two feet and had no handrails. I wasn't even moving yet, and already it was hard to keep my balance. Hesitantly, I uttered a feeble, "Giddyup."

The cultipacker lurched forwards, and I went backwards. I almost fell off but managed to hunker down and grab the side of the platform with one hand. But doing so distorted the pull of the reins in my other hand, which the horses interpreted as a left-turn signal.

"Whoa!" I screamed.

The horses dutifully stopped. I rested a minute or two, closing my eyes and waiting for my heart to stop thumping. Eventually I mustered my will, stood up, and let the dread word escape my lips again: "Giddyup." Preparing for the worst, I now planted my body on the platform in a rigid and slightly bent posture. Despite my intention to go straight through the center of the field, we were still turning left. Sharply. I tried to correct by pulling the reins far the other way. Now we were veering right. We were circling back to where we started from.

For several minutes, as I toyed with the reins, we jerked this way and that—any way but straight. The small ridges of the roller were etching most artistic patterns in the dirt. I was about to lose my patience . . . when I remembered something.

I had had similar trouble in my first lesson in driver's ed.

The teacher back in my high-school class, as my car veered in and out of its lane, had given me a tip: keep your eyes on the middle distance. Emboldened with this memory, I gazed straight ahead about thirty feet, and at once the problem was corrected. I had reconceived the team as a long, ambling front hood.

Once I got the hang of it, cultipacking became quite satisfying. At the merest tug or click of the tongue, two hirsute animals weighing over half a ton each obeyed my every command, pulling a piece of farm equipment surely as heavy as themselves. The pace was slow and relaxing. The weather was idyllic. In the stillness of the wooded enclosure, imbued by a sense of accomplishment, my mood quietly soared. Man had triumphed over beast, mind over matter, and skill over stupidity.

In a couple of hours the two-acre field was cultipacked, Sylvan's rows included. I felt pride in what I had achieved but disappointment that no one (except the horses) had witnessed it. I led the team back to the barn and turned them over to Sylvan. "How'd it go?" he asked.

"Once I got the hang of it, it was a lot of fun!"

His eyes began to roll, and I realized I probably shouldn't have acted as though the job was not work. But there was an important truth to be found here: In our new life, mechanical skills I had previously acquired were not utterly worthless. Some, like steering a vehicle, could be readapted, shifted down to a lower horsepower.

Yet now Amos stood on my back stoop, gazing up at me plaintively and tempting me to shift back up again—to the speed and conveniences of the self-pumping ram. I had to admit our outdoor hand pump, quaint as it was, had some drawbacks. The joy of lugging heavy, sloshing buckets in and out of doors was already wearing thin. Besides washing dishes with the water, we used it in the bathroom to flush the toilet. (The Minimites all used outhouses, but there was no outhouse on our premises. Not that we minded.) We also bathed in it. But the joint-wrenching act of hauling water had little to be said for it as an occasion for socializing, contemplation, or exercise.

Others around us, with all their knowledge and experience, had long ago discarded the practice. Many of our Minimite neighbors had rams. Those who didn't used gravity-fed spring water or pitcher pumps right at the kitchen sink. When Mr. Miller's married daughter saw Mary at our outdoor pump, she couldn't contain her bemusement. "How old-fashioned!" she exclaimed. Even the Minimites thought us behind the times.

Why not go with the flow?

—

I was weeding a flower bed in the side yard when I heard a ruckus. I wheeled around in time to see a flatbed wagon led by a pair of galloping horses careening down the highway in front of our house. Two figures stood atop it—Nate Jones and one of his sons. The wagon lurched this way and that and finally landed in the ditch.

I raced to the spot and found Nate and Perry dusting grass and dirt off their clothing, apparently unharmed. The horses were also okay. But the wagon was a sight. The wooden tongue had broken off. Nate had no choice but to abandon the vehicle and lead the horses home on foot.

But what had gone wrong? I knew that, to save money, Nate had put the wagon together from used materials, as he had his house. Now he told me he had never installed brakes. "Why'd I need brakes, I thought"—he said—"when I can just pull on the reins?" He'd just discovered the answer. As the wagon began to go downhill, the pressure on the horses' harness straps built from behind, pushing them forward. As they moved faster, momentum increased, heightening the pressure, and the problem snowballed. Eventually, barely able to keep ahead of the wagon, the horses panicked and broke into a run.

"If you're a beginner on the homestead," Nate bawled, "ya' take two steps backward for every three steps forward. Or maybe it's more like seven backward and eight forward. You've just got to learn patience."

The mishap gave me pause. Could the ratio of progress to regress be as low as Nate claimed? Was it inevitable that I would backslide as he did? We had made that mistake with the weeding, I knew. But this matter was different; it was a question of technology. Nate had violated a principle that I had only lately come to grasp: that too little of it is no better than too much. Maybe the beginner's error lay in viewing technology as evil in itself, overcompensating for technological tyranny with technological abstinence. If so, I was one step ahead of the game. We were about to get running water.

Amos dropped by again and asked if we wanted to see the spring. I didn't hesitate a moment. "How do we get there?" I asked.

"If you want, we could take your car," he said.

I had, of course, deemed car travel strictly off-limits except when its use was clearly justified, and even this was a temporary concession made in lieu of alternative long-term arrangements. And yet it seemed boorish to deny Amos's good-natured request. Why be uptight about a little technology? Impulsively, I reached for the keys. (At last something to show to *him*. A mere return of favors, no? A *technological reciprocation*.) To my gratification Amos's eyes bugged out. To an unworldly Minimite boy, the interior of the Escort must have looked like the cockpit of a flying saucer. I scanned the fields in case Mr. Miller was watching. The coast was clear.

We chugged up a long open hill beside a barbed wire fence. At the top, I slowed near a clump of trees. Turning to look behind us, we beheld a tiny white cottage with a diminutive garden and threadlike rows of vegetables. The oak in the front yard looked disproportionately large and dwarfed the house underneath. Carpetlike pastures and tracts of tillage rolled together in a wavy quilt for miles in all directions. All creation seemed to have unfurled and thrown itself at our feet. And here we were, regally beaming down on it from our spaceship.

We proceeded through a stand of trees that formed a windbreak down the other side of the hill, along another fence to a gate. On the other side of this gate was a field full of cattle. Amos got out and opened the gate.

The grass in the pasture was taller than the hood and lay down flat as we drove over it. The air was heavy and sultry, redolent of pollen and crushed grass. Cattle began to lope alongside us, as if we were part of the herd, masquerading as a large cow in this small vehicle. "This way?" I asked. I tried not to sound nervous. The cattle were getting a little too close for comfort.

"Pretty much," Amos answered. But he could barely see over the dashboard; he was craning his head to look.

"Are you sure," I asked a bit more pointedly, "this is all right? Are there any rocks or holes in this—" It was not necessary for Amos to reply as a horrible grinding resounded from somewhere under the car. I would have stopped, but the cattle were pressing in too close.

Nothing to do but move onward. The grinding passed, but the cattle continued to press.

"Uh," I asked, "is there a bull in that herd?"

"Yes, he's that big one running in the middle."

"But is it safe to be driving a car in a field with a bull?"

"Well, he's pretty tame for a bull. We've already worked with him a little. He won't do anythin' to you s'long's you don't bother'm. See, that kind's red'n'white, and they're the tamest. That's why we got 'im. We've had 'im a couple of weeks now." Amos had an almost British lilt to his voice that, from what I could tell, might have undulated just as pleasantly were a wall of lava descending on us. How could I tell he wasn't saying anything he could think of just to prolong the ride? Red bull. Red car . . .

I said a few Hail Marys under my breath, and we made it safely across the field to a second gate.

As the herd gathered around us, I peered beyond the opening. On the other side, a gravel road descended precipitously through thick woods. The steep, rutted lane was clearly impassable in a car. The folly of our expedition was now too clear.

I could only intone, "I really don't think we can do that."

Back the way we came, bypassing the place where I had heard the grinding sound. At least the bull left us alone. A mere half-hour had been wasted. No sweat.

But the thought of the natural water source had piqued my curiosity. The next day, abandoning the preventative hoeing I had planned, I set out again on my own to find it. This time I went by bike. The little black city bicycle would conserve a few minutes that could be used later in the garden. It had fifteen speeds and an orange flag for safety attached to a long fiberglass stick mounted to the rear axle. The bright fluorescent flag swung back and forth, brushing against the tree limbs as I crossed the gate into the first enclosure. The herd grazed in the center of the pasture, with its lord rising majestically in the midst. I felt fortunate he was so far away, doubting I would be noticed from this distance.

I literally was burrowing through the grass. I had to lift my head to avoid facial lacerations from the tall blades. Still, I enjoyed the feeling I was under camouflage, thinking my head was low enough to escape notice. Then I heard a rumbling sound. Turning my head to the left, I was startled to see cattle at my flank. They seemed to be moving rapidly and gaining on me. "Odd," I thought. "Why would they—?"

The herd was fast closing the gap, almost alongside me now, maybe a few yards away. Here came the bull, bounding and tossing his head—

My heart jumped out of my chest.

What happened next was a blur. A furious rush of pedals . . . hurling myself and my bicycle over the wire . . . gasping, doubling over . . .

I was over the fence now, but almost immediately more cattle began approaching through the trees! What? Had the herd gotten through the fence? No, I saw that it hadn't. This was *another* herd.

Cattle were nearly on top of me now, and perhaps unthinkingly, perhaps under the delusion that I might slip through them unnoticed, I jumped on the bike and headed down the steep, pitted road, the orange flag waving more vigorously than ever. The herd merely reversed its direction. Now I was at its center, a cowhand leading it onward, its flag bearer.

At the bottom of the hill, I began to cross a clearing; I shot a glance over my back.

No bull.

The herd, in fact, was breaking up, and in a minute or two, only a couple of head followed. I got off the bike to ford a small stream, then continued where the road led. But I couldn't shake them. The two head continued tailing me and—could I be imagining this?—began tossing their heads and bucking like bison, looking as if they were about to charge. I peered under their bellies.

Indeed they were male!

They weren't as large as the red and white bull we had encountered earlier; if anything, they were a bit smaller than the average. And the toss of their heads seemed casual, almost playful. They lowered their heads, snorted, then bucked again, following each other in a circle. As I continued to pedal, catching glimpses of them over my shoulder, the road came to an end, and once more I waded through grass. But then, as if reading my worst fears, one of the animals broke from its partner and headed toward me. Anticipating a five-hundred-pound wallop, I abandoned my bike. Throwing it in his path, I darted to a nearby tree. Cowering behind the trunk, I waited for what seemed forever, hearing no sound but the deafening pounding of my own heart. At last I dared to peek. The bull was a few yards away, sniffing and pawing at my bike as if puzzled by my disappearance. Without notice he raised his head and looked me in the eye. I jerked back but it was too late. The earth began to shake again . . . then stopped. Unable to contain myself, I peered around the trunk from the other side. A few feet away, the animal sniffed at some dandelions. On impulse I decided to make a run for it. The next tree was near enough, and making it, I wove from tree to tree until I had ascended deep into a wooded hillside, faintly aware of having passed a noisy fount of water gushing from a rock outcropping. Soon I was deep in the wood and had lost my bearings. I meandered towards where I imagined the farm boundary lay, for I knew that beyond it was a county gravel road. I came upon an old chimney and the remains of an old stone foundation. All I wanted was a fence to climb, something to put between me and the bulls.

Finally I reached it, scaled it, broke through some brush—and there before me sat the Millers' house. I ran over and came face to face with Amos and Caleb next to the large metal equipment shed. It seemed as though eons had passed since my last contact with human life. I poured out my story, and when I had finished the tale, I added with a hint of indignation, "You told me there was only one bull!"

Amos eyed me, incredulous. "Them's baby steers. They was just tryin' to have some fun." Amos looked at Caleb, and they laughed out loud. I couldn't believe it. I suddenly felt so angry that my face began to flush, and I had to work to disguise this and keep from saying anything I might regret. I had nearly been killed! (Hadn't I? But this was now in doubt.) Two of the older brothers had heard the commotion and came by to join in the revelry. One of them offered to get the bike back for me, but he was obviously only too pleased to say so.

As I passed Mr. Miller in the barnyard, he said (suppressing his own smile), "Don't worry about them. They've probably just been getting a big head."

The misfortunes seemed to feed on themselves. After tinkering with the ram for three days, Mr. Miller and his sons gave up. They never got it to work. It pumped well enough, but water never reached the house. They figured at last that the hill was taller than they had estimated. The next-bigger-size ram was needed, and Mr. Miller was not prepared to purchase it.

Jed returned my bicycle, the front wheel twisted like a pretzel.

The car, I learned the next time I drove it, had a new rattle and had to be taken to the shop.

I trudged back to a garden grown weedy through days of neglect, and I remembered Nate's warning.

Indeed, one tentative step forward took my pride several steps back.

the missing refrigerator

It was Mary who pointed out a second technological shortcoming of our household. I got the message indirectly.

While we were still in Boston, she had shared with me her fears about how tasks would be divided on the farm. Mary was five years older than I and had grown up in a time when many girls were expected to stay in the kitchen, while boys—for lack of sufficiently absorbing "male" work on a suburban lot—more freely pursued their own interests. And in a family of twelve children, there were only two girls old enough to lend a hand for many years. This led in turn to a further division of tasks within the female domain. Mary's sister cooked, and Mary cleaned. She didn't begrudge her parents at the time for what was, after all, a logical application of cultural norms still lingering then, but she recoiled from the thought of being arbitrarily held to the same norms today in her own home.

In discussions before arriving, we agreed that no one would "have" to do anything, that we would work together where possible and otherwise leave tasks to whomever was handiest. I assured her, most of all, that I didn't mind cooking if the duty fell to me (long hours of number-crunching and frequent dining out had not awakened in her any new desire to be in a kitchen). One of the beauties of a working household, in theory, was give-and-take: each spouse formed half of a tightly interlocking economic partnership, each provided a visible and meaningful contribution to the common enterprise. What matter the task performed, when all were essential, all were integral, to the goal? And wouldn't taking turns add variety?

I suppose it shouldn't have mattered that I didn't particularly care for cooking either. As a graduate student I had lived on macaroni and cheese and canned green beans.

Fortunately, with the exception of cooking, Mary's practical aspirations dovetailed with mine. And she had no objection to doing something traditionally considered womanly when the desire sprang from within. She gratefully left heavy lifting to me, and was delighted to find that her sedentary hobby, sewing, newly suited our needs. She was perfectly happy puttering around the house; what had bothered her in her youth was less domesticity than the feeling it had been artificially imposed upon her.

But as the summer went on and as Mary spent increasing amounts of her time at home, and I increasing amounts of mine away with the Minimites (often at one of our off-site produce patches), the person usually left near a kitchen was Mary. Edible fare on hand for a novice homesteader chef consisted almost entirely of vegetables. On rare supply trips to town we did stock up on canned meat, of which the grocery store provided three choices: Spam, chicken and dumplings, and enchiladas.

One time I got home around six p.m. after a vigorous day in the pumpkin field. I looked at the kitchen counter and saw no meal in progress. Mary was still sewing, oblivious of the time. This was not the first occasion when I had come home to an empty table, hungry. I felt a headache coming on. I cleared my throat, but Mary didn't notice. Before I knew it, I murmured, "It doesn't have to be a fancy meal. Just something before eight p.m. so I don't get a hypoglycemic attack."

Mary looked up at me from her sewing. She appeared to be coming out of a pleasant reverie.

I went on. "If we eat too late, I just lie in bed like a beached whale waiting for the food in my stomach to digest"—the pitch of my voice was higher now—"and then I don't get any sleep. And then the next morning I'm *worthless*." By the time I got to *worthless*, I realized I was shouting.

There was a strained silence. At last Mary replied, in a thin voice, "We're out of bread so I can't make sandwiches. And I haven't thought of anything else yet."

"Don't you see me dying here? Don't you have any pity? Can't you open a cookbook?"

"It usually calls for ingredients we don't have. If we could keep leftovers cool, then we wouldn't have to make three meals a day from scratch."

"What are you implying? That we get a *refrigerator*?"

I suddenly got a sinking feeling. Mary's point was well taken: our inability to cool leftovers doubled the cooking load and in doing so, fettered Mary in domestic chains she has shrunk from since childhood. The problem had no clear resolution.

The method of food preservation we had used so far was the one the Millers had taught us: canning. Canning surely had much to be said for it. It was more than a means of keeping food. It was an activity in which we worked together as husband and wife, and therefore it preserved not only the fruit of our labors but also, in some palpable way, the memory of our discussions and the reconciliation of our differences (which, come to think of it, until now had been relatively minor). The root cellar was more than a food-storage facility; it was also a kind of domestic shrine, beautiful in itself, with mauves, navies, chartreuses, maroons, canaries, violets, and siennas, all dimly glowing like stained glass.

Besides all that, the in-home storehouse was eminently convenient—a veritable grocery store a few steps from the kitchen.

But canning was no way to keep leftovers fresh for a day or two. Our neighbors had similar problems with short-term food preservation, and this was why some Amish communities had adopted propane refrigerators. By using propane, they still observed the time-honored Amish ban on electricity. But the decision was problematic. It had led to dissension in one Amish settlement on the Michigan-Indiana line. I didn't know why some Amish resisted propane, but I knew why I would. I worried whether the cost of buying and maintaining a motorized appliance would ever produce a corresponding return in convenience. I also remembered seeing lots of wasted food in the back of the refrigerator at home. And I feared gradually adding more and more gadgetry to our lives in an escalating spiral.

But this didn't necessarily rule out the possibility of some method of temporary cold storage.

Some Minimites had access to caves. Others could set their goods directly in a spring bubbling out of the ground. I thought of an economical way to simulate these natural phenomena: a smaller refrigerator powered by the sun, used only for the summer months. The solar cell would be most efficient precisely when the sun was hottest and the need greatest. It was a solution, theoretically, as natural in its own way as a cave or a spring. Most solar systems are unwieldy because long gray winters require a large bank of storage batteries. But a refrigerator wasn't needed in cold weather. A summer storage unit, on the other hand, converted the problem into the solution.

The Minimites captured naturally occurring differences in temperature using a more traditional method: by harvesting winter ice from ponds and packing it in a shed. Using sawdust saved from the lumber mill, they tightly insulated the shed's walls. Because the sawdust in the walls was so thick, and because there was so much ice inside, the ice was self-refrigerating. Hence the feat of ice cream on August evenings without an electric freezer. Merely a bit of winter trapped in a bucket. *That's* conservation.

An even more enticing way to deal with the problem was to eliminate the need for food storage in the first place. The notion had little precedent among North American Old Order groups, so it was only when I stumbled onto the works of Eliot Coleman that I learned of the idea. In *The Four-Season Harvest* he argues that the life of vegetables in the ground can be prolonged during winter by using simple cold frames and movable plastic tunnels. Coleman distinguishes his method from greenhouses requiring expensive technology and heating fuels; rather, he proposes merely extending the life of crops already planted during mid- to late summer by cheap and easy means. Thus, instead of preserving winter for use in summer, he preserves summer for use in winter. And thus, instead of canning at the hottest, busiest time of the year, one may pluck leeks, kohlrabi, carrots, and mâche fresh for the supper table right out of the ground, just when one needs them.

Admittedly, none of these methods was available to us yet. But that didn't prevent me from fancying them.

On a routine visit to the Minimite general store, Mary's eye fell on something she hadn't noticed before. Among the racks of homegrown herbs, cooking utensils, and kerosene lamp chimneys sat a cookbook written and published by the Miller daughters. She took it home and began thumbing through it. It was a comprehensive introduction to Minimite cuisine, relying mostly on locally available ingredients. She solicited some tips directly from the Millers themselves—whose resistance to her smile had by now collapsed—and began to compile more recipes from Carol and other neighbor women. She then quietly repaired to the kitchen.

Drawing from the mounting supply of summer squash we were canning, she soon presented a squash casserole, a crisp, piquantly seasoned tour de force. From the aroma alone, saliva gathered in the corners of my mouth. Corn casserole followed, timed with the arrival of corn picking. This dish was so mouthwatering that she still makes it for our dinner guests, and they always ask for the recipe (admittedly it does call for one package of Jiffy corn bread mix). Mary became quite handy with bread, and when we had non-Minimite visitors they begged for a baking lesson. Of several possible recipes for homemade ketchup, she selected one that was robust and rich, with a blend of flavors for the whole palate. Her pumpkin pie was unlike typical pumpkin pie: it was closer to chiffon.

As she began to see what possibilities the cooking craft offered, household tension eased considerably. The domestic dungeon had all but burst its bonds, and Mary the *artiste* now stepped forth.

But one thing continued to gnaw at me. It seemed a pity to throw out Mary's creations after only one tasting. And our decision to forgo electricity still constrained us from sticking food in the ol' fridge. At some point—maybe it was after our third or fourth food-spoilage discussion—a bell went off in my head.

Because of the failure of the ram, I was still pumping water from the cistern by hand. Tall buckets of it lined the kitchen wall, ready to be used to take a bath or flush the toilet. All that water sat there with nothing better to do than leak its fifty-five-degree coolness into the atmosphere.

The next chance I got, I placed Mary's food in glass jars and submerged them in the buckets using rocks to keep them under. The food lasted until the next day.

As we learned to live within our minimal means, what we discovered went beyond simplicity or economy, or getting "more for less." It was another way to be, well, cool.

the sounds of silence

Beginners though we were, I venture to say that over time we proved Nate wrong. The number of steps we took forward exceeded the backwards ones by a ratio surely greater than eight to seven, or even two to one. But whether we made progress or not on a given day, one thing was certain: the day would come to an end. Here was reward to struggles of all kinds—their simple cessation. Yet just as there was more to work than labor, there was more to rest than the absence of work.

When the dishes were done and daylight began to ebb, Mary and I would sink into the two easy chairs the Millers had placed for us in the living room and allow our muscles to uncoil. No after-dinner cordial could surpass this lovely wash of endorphins streaming out through our limbs. Light would drain from the space until little remained but a gray gauze. At the flick of a match, the house lights would come on, and then the music would begin.

At first the sound blended in with the general lushness of dusk, and we unthinkingly reclined in it as we did upon our chairs. Gradually becoming more distinct, then, through the tall open windows and the proscenium formed by the columns of the front porch, it would roll in and wash over us, wave after audible wave. Soon we would be tingling crown to toe in a sympathetic vibration as if our bodies were musical instruments themselves (which in a sense they were). Hundreds of unseen, unpaid fiddlers were bending to their nightly duty.

Almost as soon as we took note, the music would break off.

Crack! Nothing. Perhaps a twig had snapped, or some hacker back in the seconds had skipped a beat, throwing off the timing and triggering confusion among the players. Or perhaps they had somehow become aware that they were being heard. Whatever the cause, the music stopped. A lone owl could be heard hooting as if on a pitch pipe, to no avail. A distant dog would bark like a coughing spectator.

Then silence would descend.

It could go on for minutes at a time.

When had we had silence like this? The music of the crickets was one thing, but the silence was everything, all-enveloping. It was almost disorienting. In the twilight, you had the feeling the room was slowly spinning. When the vertigo passed, the silence fostered a calm deeper than any sound.

As we settled into our easy chairs, lulled into a contemplative mood by the quiet ebb and flow of sound, we turned to our books. My selection for the night, *The Education of Henry Adams,* was a tome that had long daunted me. It had been assigned in two courses, one at the undergraduate level and one in graduate school. The autobiography of John Quincy Adams's grandson and a man of letters, it had set the tone for a whole era of Americans. It broached the subject of technological infatuation in our nation's past. And I had never been able to get past the first two pages.

Tonight, to the flicker of a kerosene lamp, I made inexplicable, rapid progress. When I got to the twenty-fifth chapter, "The Dynamo and the Virgin," I found Adams falling prostrate before the dynamo at the Great Exposition held in Paris in 1900, a huge electric generator with a giant cranking arm. To him the slowly undulating device symbolized a new and unprecedented Force that supplanted once and for all the animate energies of nature. A narrow and inhuman power at last had vanquished Fecundity, or Reproduction, which Adams personified in the Virgin and portrayed as the source of natural bounty. But even as he wrote wistfully about her passing, he bowed before what had replaced her.

Adams's poignant shift in allegiances haunted me, and I couldn't get enough of his writing. Why was the book suddenly so clear and full of insight to me? The subject matter was largely autobiographical, and not directly related to my field of interest.

In the modern university, with its rapid turnover of assignments and fast-paced technology, the human brain is treated as just another processing device and is expected to keep pace with electronic blips. But Adams's thought, ponderous and discursive as it was, could not be summarily ingested. He had lived within a culture whose movements were still largely limited by the speed of horses; the ambling cadences of his writing preserved this pace. Having taught medieval history at Harvard besides, his verbal nuances hearkened back to an earlier epoch still and seemed to echo from the deep wells of time, the vaults of the great cathedrals.

This was the secret: to grasp his meaning, you had to be living it. Not merely your thoughts, but your various daily duties, the material accoutrements by which you performed them, had to fold together in a quiet rhythm, an interconnected unity.

And this explained not only why time moved more slowly but also why we had more *of* it, why we were able to relax and read the way we were doing right now: in the absence of fast-paced gizmos, ringing phones, alarm clocks, television, radios, and cars, we could simply take our time. In being slower, time is more capacious. The event is only in the moment. By speeding through life with technology, you reduce what any given moment can hold. By slowing down, you expand it.

Shortcuts lead to emergency mending sessions in order to piece back in what was cut out, to lengthen what was shortened: Computer users, cramped in a cubicle all day long, jogging around the block. Bureaucrats and financiers, zooming ahead along their career paths, then reversing gears to attend school concerts, ball games, and parent meetings. Captives of the technological environment fleeing for brief weekends to mountains, beaches, and rustic cabins.

What began as short lines become circles—myriad overlapping loops that, described on paper, resemble nothing so much as the cloverleafs of our freeway systems. These roundabout routes to satisfaction leave their followers less time than ever. For the better part of the day they are in transit, simply speeding forward, never arriving. In a world in which everything is in motion, motion finally comes to seem the absolute, the unfailing standard by which everything else is gauged. Progress becomes its own self-justifying logical loop, the endless cycle like that of the dynamo before which Adams bowed. Only from somewhere outside the loop, only from a position of true stasis, could it even be noticed. Only in the deep repose of a summer's evening, serenely planted in a twilit cottage, savoring rich ideas, could I regain my bearings.

As I whiled away my time with the grandson of a president, Mary sat beside me, arm in arm with the Spanish mystic Mary of Agreda.

———

I was still deep in my book when, in the corner of my eye, a spark flashed. There, at the other end of the room near the ceiling, another spark flashed, and another.

"It's a firefly," Mary observed. We peered over at the little flicker bobbing up and down, and a spell fell on us. We eventually roused ourselves and returned to our books, feeling snugger knowing a little fairy was guarding our cornice.

What drew him in? Could it have been the kerosene lamp? Was there something amorous in the fly's infiltration? Or was it simply that he had mistaken our room for the darkness outside?

Either way, the speculation was intriguing. It suggested that, just as the cessation of mechanical noise opened the air to soothing sounds and the removal of technological efficiency made way for natural rhythms, so did the absence of artificial glare invite the play of natural illumination—a performance in light. Yet here again, the phenomenon was not merely an absence but a presence.

We had never noticed stars in Boston, but now they spread across the sky in their full celestial glory. The firefly was but a small star, in terms of brightness, and when we opened the door and followed it outside, this star was released to a galaxy of stars, into which it blended almost imperceptibly. Starlight, when congregated, was enough to light your way to the garden for a midnight snack. Imagine opening your kitchen cupboard and grabbing a handful of cookies under the glow of two billion stars.

Stars were not the only handy visual aids. Early in the morning, accompanied by the lauds of the rooster, another made its appearance. Pale at first, slanting across the floor and spreading along our sheets, then pressing against our closed eyelids, it roused us from sleep almost as gently as starlight had lulled us into it.

Kerosene lamps were useful not so much for what they did as for what they didn't do: obliterate a system of timed illumination nature long ago instituted, perfectly good for lulling us to sleep and waking us up. (Researchers have discovered that the sun activates a chemical in the brain to rouse us.) At most these demure sources of light took nature's circadian idea and embellished upon it. A glowing oil lamp is but a vessel of concentrated starlight; its flicker but a domesticated firefly. At best it can prolong natural light but a little. When one's eyes begin to dry up and one's lids to droop, nature takes over. The bedtime cue has prevailed. (I will admit that sometimes I cheated a little by setting a mirror up next to my lamp to double its output, but I paid the price when the sun came out on schedule the next morning.)

The Minimites, we found, supplemented their kerosene lamps with flashlights. We were surprised at first. But in time, we too saw the need. Without robbing us of the magic and economy of kerosene, they proved handy in the middle of the night when you needed to get up quickly for any reason: going to the bathroom, checking out a strange noise on the porch (some noises were stranger than others), tending a wailing animal. From a distance a farmer carrying a flashlight in the darkness looks like a firefly.

Mary and I, in any event, preferred the sparks of an amorous firefly to those of a shorting fusebox—or nervous system.

And lolling by the light of the flickering lamp, I relished one additional perception: here was another contribution to leisure. Now that we had so much more time, the built-in cycle of daylight and dusk saved us some of the chore of managing it.

The signs of an elegant fit between the natural order and human well-being revealed themselves fleetingly, then faded into the background. You had to pay attention or you'd overlook them. In truth, it was all too tempting to enjoy the effects without acknowledging the source. In the dreamy world of twilight, it was even possible to think you were imagining things. It might have been that our machine-starved minds were hallucinating.

But if so, we were not alone. Occasionally our neighbor, Mr. Miller, used odd terms that at first I didn't catch, like "fast time" and "slow time." Whenever I set an appointment with him, or inquired after the time, I had to remember to specify which kind I meant. I learned that "fast time" referred to the modern convention of "daylight savings," whereby workers are artificially ceded an extra hour of daylight in the evening, after subtracting an unnoticed hour in the early morning. But when our landlord said "fast time," I could detect in his throat-clearing a certain contempt the phrase aroused in him: the folly of man-made time. "Slow time"—the preservation of the natural markers of dawn, noon, and dusk—was his response to the harried chasing after artificial objectives. "Slow time" then did not merely refer to a different number on a clock; it was the Minimite acknowledgment of an entirely different structure in life, an entirely different pulse. Leisure didn't end when work began, but pervaded every moment of the day.

growing

quickening

As the season progressed, to be sure, there was a discernible quickening in the tempo of activities. Natural time, like light, was not constant but varied according to natural cycles.

At the sun's gentle insistence each morning, I'd rise at about six-thirty (albeit a little earlier each day until June 21) and stroll over to the barn to milk the cow. The session that at first had produced less than a gallon eventually yielded almost two. Besides increasing the workload, this surge created another challenge.

What would we do with the milk? We didn't have a refrigerator, but even if we had, there was too much to think of drinking. "Cottage cheese," said Mrs. Miller. "Cream cheese," she said. "Butter. Cup cheese. Mozzarella. Monterey Jack. Custard." Soon Mary had a cookbook open to the cheese section and was boiling vats of milk, hanging socks full of curds from the porch eave, and setting out platters of white ooze to cure. The cottage cheese was easiest and seemed to taste the best.

And still we couldn't keep up. We were pouring milk into the ground. "You need a pig to drink that up," somebody said. A pig?

But it was the height of the summer season, and growth and activity were beginning to fill the available space and time. We were busy gardening, zipping up to the Millers' to tend the pumpkin patch, traversing the community to Sylvan's to mind the sorghum, and brooding over the canner in the kitchen, even as the Millers continued to entice us with other new opportunities-cum-obligations. We were bent over with armloads of fresh corn, beans, cabbage, and potatoes; we tugged out more surly weeds (hitching up garments that no longer fit us as our waistlines slimmed); cracked open chestfuls of chestnuts from the trees around the garden; learned about grapes and grape jelly; (stealing a few minutes here and there for those sweet, private moments all newlyweds know); loaded up endless jars in the pressure canner; then sorted, labeled, boxed, and—triumphantly— shelved the finished goods in the root cellar (not sure if we loved the homesteading because we loved each other or vice versa).

—

Knock, knock.

"I can't believe I forgot to shut the bedroom door."

Mary looked up in surprise.

"Oops."

From our bed it was a straight line through the kitchen to the back entrance. And there, peering over the sill of the door, shaded by the brim of a straw hat, were two curious eyes. It was the middle of the afternoon. Mary and I were clothed only in sheets.

Moments later, and a little disheveled, I once again greeted Amos, our regular visitor and farming consultant.

With his slightly beseeching, singsong voice, he made a polite inquiry: "Father wants to know if you'd like to get some apples. There's a man across the county who needs some help with his, and you get to pick up as many off the ground as you can take."

"Well, I don't know. Maybe, but we're kinda . . . busy right now."

"It'd be for next week sometime."

"Why don't we wait and see, and maybe we'll have the chance to do it."

"That's up to you." He turned to go, then hesitated. "By the way, we fixed the hoe."

One day I heard a muffled sound in the living room and went to the doorway. There was Mary hunched over the armrest of the easy chair, blowing her nose. Were her allergies acting up? She had been taking medication for this.

When she raised her head I realized her eyes were too red for sinus troubles. She was crying.

"This is silly," she said between sniffles. "It makes no sense. But I miss my mother."

"Your mother? But you hardly ever saw her even when we were in Boston."

"I know. It makes no sense."

I pondered this strange turn for a few days. At the same time Mary began to grow unaccountably tired. She was losing enthusiasm for gardening. She wanted to sleep in. She also noticed she was late with her period. . .

—

Now we were in deep.

At first Mary panicked a little. The Minimites all had close relatives in the immediate vicinity who tutored their young mothers and fathers through the delicate pre- and postnatal changes. We didn't. That was why morning sickness had triggered homesickness for relatives and old friends. But there was more to it than this. Mary admitted missing going to church. There was no recognizable center of Catholic life in this area, and for all the problems the church was going through, she felt a need to grasp hold of the familiar and the solid.

I panicked a little at Mary's panic. It was as if she suddenly had drawn a blank on everything we'd learned and experienced here. The fear began to eat at me that she might be unable to last out the eighteen-month commitment. To me, this was unthinkable. To head off a possible disaster, I had an inspiration. I proposed a diversion: to take a little scouting trip to places where we might live once the year-and-a-half stint was over. As we had agreed, of course, that at that time it would be Mary's turn to pick our home.

Luckily the sorghum was tall and the pumpkin leaves thick, so it was barely necessary to weed them anymore. The canning was done and the rains had abated. My beard had almost filled in. There was hardly anything to do. The quickening had crested, and we were in the midsummer lull. It was the perfect time to get away. I had resolved to use the Escort only if truly necessary, but I thought, given our special conditions and requirements, the time had come.

I took one parting look at the pumpkin patch. With diligent hoeing and regular rain, the vines had all crept together to form an impenetrable green tangle. Were any actual pumpkins growing under all those flopping tendrils? I lifted my leg, thrust it through the leafy canopy, and set it down gingerly, toe first. Repeating this maneuver several times and moving toward the interior of the field, I finally spotted something interesting. What was it? Was the green bulge under the yellow blossom a cucurbit wart, a symptom of some dread disease? I bent closer. No, it seemed too evenly rounded and striped to be pathological. My heartbeat quickened. I took a few more steps, then stubbed my toe on something bigger. It was large, round, and green, with ridges like a huge grenade. I drew back and parted the leaves.

It was six inches, no, eight inches across! After only a month? I gazed at it with the adoration of a father beholding his firstborn child.

—

The first stage of our search took us east. After further discussions, Mary and I had worked up a short wish list: (1) Close to friends and/or relatives; (2) Church in vicinity; (3) Walk or bike to centers of life; (4) Amish nearby; (5) College (as an employment possibility for someone with a postgraduate degree; I had little desire to join in the race for tenure or prestige, but I didn't mind the thought of teaching part-time at a quiet liberal arts school). We realized it might be hard to find a place that would fit all the criteria, but meeting some was better than none. The hunt for a nice college town led us to Steubenville, Ohio.

Steubenville was the home of Franciscan University, a small Catholic school lately enjoying a resurgence. Several friends of ours were affiliates of the institution. There was even an Amish settlement not too far away, which we hoped might provide some tie to an agrarian existence. Steubenville seemed to offer a little bit of everything we were looking for.

As soon as we arrived, our hearts sank. The former steel-manufacturing municipality on the upper Ohio River, across from West Virginia, looked down-at-the-heels. Where a bustling commercial district must once have been, near the water, was a collection of mostly derelict brick buildings. Worse, this bare and depressed-looking downtown sat isolated from the rest of the community, which was scattered in clumps high on the hills overlooking the industrial river. You could get from river to hill only by car. Nor was the university well situated with respect to the rest of the town, but again on its separate hilltop site that could be approached only by automobile. The college architecture was modern and functional, built obviously with a mind to cost savings above all else. Everyone we came in contact with was friendly and outgoing but openly bewailed the ugliness of the environment. It seemed to lack a certain minimal human appeal or accessibility.

Moving eastward, we came to the next stop on our journey. Front Royal, Virginia, a town about sixty miles from Washington, D. C., was home to Christendom College, another school that offered the prospect of a ready-made social life. Several of our friends had graduated from Christendom, and the names of some of the faculty were familiar to us. But it would have been a rather closed circle because Christendom, like the Franciscan school, was isolated from the surrounding community. In fact, it wasn't really in Front Royal. It was out in the country, miles away on a twisting back road too narrow for anything but motor travel. Not that anyone in the town appeared to walk anywhere. Aside from a rather nondescript huddle of commercial buildings and a few traffic lights, there was no sign of a true center or of face-to-face human contact.

The college, admittedly, was different. Its architecture was traditional and inviting, the campus intimate and pastoral on bluffs high over the Shenandoah River. However, the isolation from the surrounding community, or what passed for a community, was troubling. Its location again presupposed use of the car as a primary form of transportation, dooming its associates to technological dependency.

It was perhaps poetic justice that, in opting to take a car on this hunt, it led us only to places amenable to cars.

Beyond Virginia the terrain up the coast towards Boston grew ever more expensive and congested. After quickly canvassing a few other places no less problematic, we turned around for home.

Home. Viewed from afar, our residence next door to the Millers took on a new and nostalgic meaning. After comparing it to some of the alternatives, we had to admit it was looking better than when we had left.

The diversion succeeded beyond my expectations. The compass of Mary's homesickness reversed its arrow.

a church meeting

Perhaps our flight from the Minimites had been too quick, too single-minded—this might explain how little it achieved. Until this point we had been rather ambling, if not aimless, in our daily course of duties. Not a very fruitful route, you would have thought, except that the thing we ultimately sought, leisure, was itself less a movement than a kind of centering. Its motions were slow and imperceptible, like growth, gradually enlarging from the core, quietly drinking in one's surroundings and commingling with other growing beings, joining together in forms of mutual nourishment. Given what it was, perhaps, the slower you moved, the more you achieved.

This ambling approach did not easily lead to instant end-results. There was no button you could push to get them, no device you could procure to hasten their manufacture. I couldn't help wondering if the places we visited, though, suffered from this very wish on the part of their inhabitants—to bottle peace and quiet, to package pastoralism. In particular Front Royal, the last haven for Washingtonians fleeing the big city, was devoid of a sense of cohesion or place. As the famous line goes, when we got there, there was no there, there. The sixty-mile-per-hour search for a nice place to live had nipped its possibility in the bud.

One of the great comforts of our present home, on the other hand, was that we hadn't chosen it. It had chosen us. Inevitability is surely part of the sense of home, the irreplaceable something for which the heart yearns. We fell through a doorway from a place we never knew into one we didn't choose. We yield irresistibly to forces beyond our control. The result is not merely a place to live; it is who we are, a deep and abiding mystery in the formation of self that can never be fully unraveled.

Admittedly, upon first arriving, our expectations had not been too high. We had tread lightly, hoping to steal our way into what appeared to be a closed cultural enclave. But now that we were in its midst, the surprise was on us. Even more obliquely, almost like the pumpkin vines, the Minimites were angling and ambling their way into our habits of mind and being. Our clothing, now half-Minimite, prompted us to wonder: had we pulled the cloth over them, or had they pulled it over us? Our hosts had perhaps one advantage: their beguilement was unintended.

One of their favorite sayings was "Do not let the right hand know what the left hand is doing." Another was "He who seeks to save his life loses it; and he who gives up his life saves it." These adages, of course, came from the Bible, and they gave expression to the disposition the Minimites held chief among Christian attitudes, *Gelassenheit,* or self-surrender. *Gelassenheit* referred less to any particular aim than to acceptance of what may be, a larger and partly hidden design that they did not fully understand.

Modern technology, I suspect, far from being neutral in its effects, has more than one underlying purpose or built-in tendency: besides reducing the need for physical effort (a kind of material surrender), it helps us avoid the need for cooperation or social flexibility (a kind of social or metaphysical surrender). All too readily it countermands the uncertainty that goes with *Gelassenheit*. Cars, telephones, message machines, caller ID, and e-mail grant us unprecedented powers to associate with whom we want, when we want, to the degree we want, under the terms we want, finessing and filtering out those we don't want—and thin out the possibilities of social growth accordingly.

Mary and I had delayed paying a visit to our neighbors' church, I suppose because we feared deeper social entanglements. Why wasn't there a "delete" button we could push to eliminate this part of our exploration? We dreaded an awkward theological standoff. Long ago, certain of our Catholic predecessors had punished Anabaptists for the crime of adult rebaptism by burning them at the stake or flaying them alive. The Minimites had not forgotten. Vivid reminders filled the pages of a book that rested on a prominent shelf in nearly every Minimite parlor, the *Martyr's Mirror*. Our only defense was that Catholic political authorities had not had a monopoly on cruelty during the Reformation, that Protestants had been no less ferocious in their persecution of Anabaptists. Not that this was cause for bragging.

Still, were we to continue our relationship with our neighbors, there was only one logical next step.

Feeling like mangy goats approaching a den of wary sheep, we pedaled toward a low and nondescript meetinghouse that was used as a school on other days. Today, Sunday, it would serve as a place of worship. Inside, hoping no one would notice us, we tiptoed into a large, bare room crowded with Minimites in plain dress. As was traditional among Old Order groups, men and women sat apart. After squeezing my way through the rows and seating myself gingerly on one of the hard wooden benches, I cast a guilty glance around. On all sides living hillocks surrounded me, broad shoulders hunched and rippling in dark blue or brown long-sleeved shirts draped in blue denim vests. I winced to reflect that I alone wore a short-sleeved white shirt without a vest. Across the room, amid a flock of white head coverings, Mary was the odd bird with the purple kerchief tied over her hair. The men had all left their hats at the door, and for the first time I noticed the white band that spread across each forehead where the sun hadn't touched, a halo earned from a week's work out-of-doors. I had taken my hat off too, and from the eyebrows up must have appeared no less holy.

There was a hush. Bishop Henry and three other bearded men moved to the front of the room and took their places on a short bench facing the congregation.

A lengthy silence followed. Nervously my eyes darted around the room and out the window at the trees streaming with morning sunlight. At last, someone in the congregation called out a page number and intoned a pitch. On this cue hymnals opened and the room began to rock. What was happening? Was it music? Chant? The sound was thunderous, yet at the same time high and thin. The slow rising and falling of tones, the curious tumbling ornamentation, offered little in the way of melody. Yet there was something subtle or supple about it: I thought I detected quarter-tone shifts, and they were executed in flawless unison. The effect was of a carefully choreographed whine.

Why such a mournful sound? Why weren't the Minimites rejoicing? Wouldn't the cumulative act in a life of harmony be harmony? One scholar, I later learned, had alleged that the type of song I was hearing had descended from sixteenth-century monastic chant; other scholars had countered that this chant had independent origins.

Either way, there was a long history behind the music, a history of precarious Anabaptist fortunes among religious enemies. Present company could not escape the memory of their predecessors' suffering any more than they could ignore their uneasy relationship to the world around them today, a world that challenged them on almost every front. This strange and melancholy chant perfectly expressed that discomfiture. The music eerily wavered somewhere between the harmonious and the off-key, yawning over the edge of tonality one minute, circling back for a resolution the next. The Minimites were not tone-deaf. The course they charted among the hazards and temptations of modern society created dissonance.

Yet what the singers remembered was no less important than what they forgot: themselves. This music reached for something higher. And it approached that something circuitously, weaving and dipping, querying and importuning, certain nonetheless of what it sought: the Almighty, the Supreme Being to whom creation owed its existence. Despite the vagaries of human life, God existed, but his ways were not obvious. This avowal conceded neither the trepidation of a supplicant before a tyrant nor despair at a being inaccessibly remote, but rather the unflagging persistence of the searcher into mystery, consoled even in partial reconciliation or ambiguity. Something inside me softened. I knew German passably and soon found myself, without really meaning to, singing along softly, following the Gothic script in the hymnal someone passed to me.

O Gott Vater wir loben dich
Und deine Güte preisen;
Dass du uns O Herr Gnädichlich
An uns neun hast beweisen
Und hast uns Herr zusammen gführt,
Uns zu ermahnen durch dein Wort,
Gib uns Genad zu diesem.

[O Father God, we praise you
And cherish your gifts;
Give us grace
That they be revealed unto us,
And that we be led together,
Admonished by thy word.]

The one hymn lasted over ten minutes.

During the course of many verses, my voice rose and blended imperceptibly with the congregation's until they were hardly distinguishable.

When it was over, Bishop Henry stood up and began to speak. He droned on like an auctioneer, first in Pennsylvania Dutch and later in English translation for the benefit of newer members and visitors like myself. It was not a very interesting sermon, and my mind began to wander. Dimly I caught bits of a message that was as banal as it was repetitive: "stay to the way"; "don't deviate"; "follow Christ"; "seek a heavenly reward"; "shun the flesh." Despite the monotony, there was something soothing in the sermon, something pleasant and narcotic in its very repetitiousness. I realized later that this may not have been entirely accidental. The effect was literally hypnotic. As the repeated phrases ebbed and flowed, I seemed to float from my seat and tumble dreamily with the others in the room who were doing the same. The sermon was like an incantation blending the congregation together into a collective altered consciousness.

At some point I heard a sound like the low putt-putt of an outboard motor. I looked and saw, three rows over, a man with a gray beard, snoring. His head was tilted forward on his chest, rising in rhythm with his breathing. Then something even more extraordinary began to happen. Other heads began rolling forward and popping back up again. Soon the whole men's section was like a shooting gallery, with moving targets challenging the expert marksman. It was a contagion from which I could not escape. I tried propping up my sleepy skull with my right arm, resting in turn on a crossed left leg, until the sweat from my palm greased the downward slide of my cheek. *Up* popped the head. *Ouch.* It was really an unpleasant involuntary reflex, as if God himself were spanking me inside the brain.

Next I tried holding my head erect and sleeping with one eye open. This was about as effective as trying to breathe with only one nostril, and soon enough—*up* popped the head. Ouch. What was the use. Not even weeding thistles had been as physically challenging. I doubted a gung-ho Marine inured to sleep deprivation could pass this test. I was in the row next to the wall, so I leaned my head and shoulders back and—

There was an interruption. A child began wailing in the women's section and the mother, unable to quiet it, quickly marched it past the whole assembly and out the door. Through the open windows, all heard the ensuing spanks. Spank, spank, spank, spank—(we were all counting)—spank, spank, spank, spank, spank, spank.

Ten. Ten was the number we had all arrived at. It was the reverse of counting sheep, and it did the trick. Though at the expense of the miserable child, everyone was wide awake. When the mother came in, she and the red-faced daughter looked wretched, but we were refreshed and indebted to them for the diversion.

What about this practice of spanking cranky children in church? I didn't dwell on it at the time, but later I reflected a little. For the most part, these young ones are supremely well behaved. The reason is manifest. In the strict German tradition of child-rearing, which these immigrant-descendants retain, disobedience to parental will is simply not tolerated. The slightest lapse or transgression is roundly rewarded, so it takes only a few lessons before the child wises up. This unflinching submission to authority may help to explain why even Amish adults submit meekly to the regimentation of Old Order groups. But even the sturdiest willpower cannot withstand certain challenges—such as staying awake during a boring church service.

After about an hour, Bishop Henry's monotone trailed off. One of the older gray-bearded men, presumably the deacon, stood and read the Gospel according to Luke, Chapter 14. He read it first in German, then in English.

It was the passage where Jesus told the parable of a certain man who planned a banquet. He invited many guests but they all turned him down, citing various excuses: one was buying land, another proving oxen, another doting on his new wife. At this the host grew angry and shifted his attentions to the poor, the crippled, the blind, and the lame, inviting them in place of the others, who he vowed would have "no taste of my supper."

On the surface, the scripture appeared to be an open exhortation to enjoy life, to feast and be merry, to set aside necessary work and join in a party thrown for no apparent purpose other than the host's desire to hold one! I was curious to see how the puritan preacher would interpret this.

As the congregation settled back in their benches, Henry's son James, who was one of the assistant ministers, stood up. He was much younger than the others, with a dark black beard and an animated face. Fire danced in his pupils. Turning a wrathful gaze my direction— I suddenly wanted to shrink back, but there was no place to retreat to—he began speaking in a deep and gravelly voice. Eventually he came to the English version:

"I hope you have come with the right preparation. I know many of you have. If we are thankful for God's many gifts, we will be filled. Filled with the word of God, the bread of life.

"If we are hungry in a natural sense, we don't need sugar coating on our bread. The bread alone will satisfy us. But most of us never have gone hungry for natural food. So we often want to dress it up and make it fancy.

"If we have prepared room in our hearts, we will savor the taste of the Gospel. Like natural food when we are hungry, it will fill us with thankfulness. But if we haven't prepared, over time we will lose our taste and pursue vain amusements.

"And when we come to trials, tribulations, and temptations, we will lose heart. These we can endure only by a willingness to suffer them. If we are hungry for food, we can also hunger for even these hardships, and discover Christ's meaning when he says, 'My cross is easy and my burden light.' If we avoid the cross, we follow two masters; we place a severe burden on ourselves. We are of two leanings.

"Christ leads us only along the straight and narrow path. Some may try to take a detour and still hope to reach the same destination."

This added a necessary corrective to a too-literal interpretation of the passage. The banquet was not really a carnal festivity, but a feast of good gifts that come to us when we least expect them, a common life that transcends the individual self. To be able to partake, one must prepare, one must learn to savor the taste of surprise, to cultivate an openness to realities beyond ordinary certainties. These higher gifts, in the Minimite view, only the Gospel could teach and only God could ultimately supply. Minister James acknowledged that there indeed were difficulties to surmount before we could enjoy spiritual festivity, but the main one was our own too-stubborn will. Its only antidote was *willingness*. The word, I realized, was probably a rough translation of the Anabaptist term *Gelassenheit*.

The fire of the preaching never relented, and when the service ended after another lengthy hymn, I felt refreshed, even purified. Returning to ordinary consciousness, I joined the men around me, who were standing up and shuffling toward the exit with rejuvenated purpose, or unpurposefulness. Having endured the greatest physical and mental test of the week, we poured out-of-doors into the light of midday, ready for an afternoon of feasting, quiet conversation, and rest. I moved through small groups of men and shook hands. I could see Mary a few yards away, talking with the women. No, they weren't ostracizing us, even though we were outsiders. Minister James, after all, had urged them to include the blind and the lame.

The service had not been a waste. In its solemnity, its chanting cadences and eerie beauty, it rivaled the most memorable Catholic ones we'd attended.

—

Several days passed before the glow from the meeting faded. It was easier to see why so many of these people had become converts. If they suffered together in the fields and if that suffering changed into conviviality, then they suffered even more in the meetinghouse, and this ordeal blossomed into even greater mutual relief. In all seriousness, the preaching about self-surrender hit home. It aroused a real ache. I ached for something missing in my own religion, something it once aspired to supply—to join human, God, Earth, and spirit in what was called "the great chain of being." I yearned to be part of a greater whole. I had left the Lutheran tradition of my upbringing because it pronounced the human sphere intrinsically defiled, and I joined the Catholic church because it endeavored to christen that same sphere and call it good—not merely to cover over but to repair and transform the damage wreaked by human choices. But the Catholic totality was now in tatters.

Yet, to enter *this* company . . . Even Mr. Miller held back. The core group of believers, he informed me, had taken the concept of a seamless totality one degree too far and pronounced it not only good but also the exclusive possession of this small enclave. There was only one church they acknowledged to be true, and this, plus a small sister community farther south, was it. At least the Catholic church allowed for a variety of spiritual expressions and communal orders within its ranks, however ragtag they had become.

Mr. Miller, at the same time, found a way to get along here without being a member. He wasn't the only one.

But to do so he and his family, and the others, had made certain concessions. Most conspicuously, they had adopted the local code of dress. For the Millers this was not too drastic a change because they had followed a similar one in Pennsylvania. For us, though, the matter was different. The onus fell particularly on the woman.

When I gazed at a Minimite matron's white, pleated cylindrical head covering, I saw a small barn silo overturned on her head. Was she overweight or not? This you would never know, given the bulky apron that drapes, like a piece of chain mail, over the midriff. Her eyeglasses are small, dark, and round, like those of a gemologist. She can see you, but you can't see her. In public she wears a thick black bonnet with tie-strings at the neck that wiggle like spider legs. Like the viceroy moth, which mimics the colorings of the bitter-tasting monarch butterfly, she thus renders herself visually inedible to the potential predator. To modern outsiders like us, such bodily bulwarks appeared to have been inspired by a puritanical desire to bind women in protective armor and keep them in their place. It seemed stilted to adhere to a fashion of dress from which everyone else had desisted a hundred years earlier.

For a while, Mary had made at least a modest attempt to abide by local sensibilities. We knew that core members of the community considered any dress higher than ankle-length a mini-skirt, and they recoiled from women in pants. So whenever she was out in the open, she wore a long dress that buttoned all the way to the neck. She didn't keep up the practice. An adventurous grasshopper got under the folds of her skirt while she was gardening and ripped away her sense of modesty.

"Modest? That's not modest!" she cried. Then and there, she threw caution to the wind and changed into her pants.

Mary now wore trousers in full view of the neighbors buggying by. This exhibitionism had gone on for several weeks. Was she bringing scandal? She couldn't tell if the women had grown cooler towards her, or if the occasional pauses in conversation were the reticences that had always been there. Finally she asked.

"Immodest?" replied Sally, an older daughter of one of the neighboring farmers. "Oh, we've never been bothered by what you wear. For you, pants are modest. It's people like Betsy or Clara that shouldn't be wearing pants." Betsy and Clara were two neighborhood maidens who were, shall we say, heavyset. What was this?

To Sally it wasn't the svelte that were worrisome; it was the rotund. Thus Mary inferred that it wasn't so much the looker who was being protected from lustful thoughts; it was the instigator of looks who was being buffered from embarrassment to herself and the witnesses. This threw her for a loop. If Sally's feelings were any indication of the dress code's intent, it wasn't puritanism but something almost the opposite. Rather than compelling everyone to be the same, it was a means of embracing those who were different. Sally wanted to save the fat from becoming marginalized.

We had gotten it all backwards. Suddenly we realized that Minimite rules on dress shared something with other Minimite rules, like those on men's beards, or for that matter on technology.

Mary returned to the garden in pants, her conscience clear. And she decided to continue wearing a long skirt when she visited other Minimite households, having grasped better the rationale for doing so.

With this misunderstanding removed, it became easier than ever to mix in, and before long Mary began to wonder whether the friendships she was striking here weren't as deep as any she'd had in Boston. People are people, first and foremost. The fact that they were not Catholic, and that we would never be Anabaptist, seemed less and less material.

a barn raising

As Mary moved deeper into women's circles, I remained somewhat isolated from the sphere of Minimite masculinity. The church meeting had helped greatly, but in some ways it also brought anguish by reminding me of our separateness. In addition, it introduced many unfamiliar faces, many identities yet to be revealed. I still knew next to nothing about most of my neighbors. They shared certain attributes with the Amish, but Amish they were not. I had little inkling even now of where they came from, what they were after, or who they really were.

The news reached me of a barn raising about to take place on a farm somewhere in the community. Lightning (in the figurative sense) had struck. A "calamity simulation" such as this might be the perfect chance to make better acquaintance. I decided to put in an appearance. To blend in better, I wore an Amish-style straw hat.

My first barn raising, I later realized, was a bit atypical. Most raisings take place in the fall, when the crops are in and the weather is cool. The barn-builder here was too impatient to wait four months and bumped his raising up to the midsummer lull. The temperature and humidity that day, I hate to say, hardly evoked feelings of festivity. As I came within earshot of the white frame house across from the building site, I chanced to overhear heated words—and I mean heated—floating in my direction.

"You'd better never have a work bee in this weather again," cried a female voice. "So much cooking in this heat."

"I'd gladly trade you," said a male voice deferentially.

"I would too," returned the indignant female, her voice rising, "but I *don't know a thing about your work.*"

"I'd come in and make the whole meal," vowed the male, secure in the knowledge he wouldn't have to.

It was in some ways comforting to know that Mary and I were not the only couple to fight in the kitchen. Minimites evidently did not step right off a postcard; the argument made me feel closer to them than all the nice talk in the world. Come to think of it, the fact that they aired their grievances so openly in English indicated they were transplants. I later found out that Minimite converts were not exempt from the need of marriage counseling.

But there was another aspect of this event that departed from a standard barn-raising format: the barn had already been raised several years earlier. Today we were only adding a fifteen-foot extension to the handsome, steeply peaked gray structure that dominated the homestead. The customary numberless horde would not descend, but only a handful of neighbors. I counted about twenty men and two women by day's end.

The scene that now greeted my eyes was nonetheless spirited. Minimite men buzzed about the existing barn, pulling boards from piles of lumber, carrying them here and there. Clad in solid colors and straw hats like mine, their shirts dark and gleaming with sweat, they chatted and heaved and toted. Small children scampered around the periphery, unable to contain their glee, pausing to watch the activity to which the men bent.

Some worked on the ground, trimming wooden posts with bucksaws; others cut sheet metal and toted lumber; still others laid block.

My gaze moved first to the sawyers. From several large piles of boards, they were selecting two-by-fours, two-by-sixes, and six-by-sixes. After laying the boards across sawhorses and marking the length to be cut, they would brandish their sharpened hand tools and trim the narrower pieces down to size in fifteen to twenty seconds. The thicker six-by-sixes took two workers several minutes on a flopping bucksaw.

Not every pair was equally efficient. Bucksawing, I soon came to understand, was a low-tech matchmaker of temperaments. Every man had a different idea of how fast the blade should move, but like rowers, each had to learn to meet the other halfway and heave at the same beat.

Nearby a stonemason was laying concrete block for horse stalls. It took me a while to make out his rapid movements. First he would lift, with the whisk of his trowel, a bladeful of mortar from a hawk board. Then, setting the dollop on a block, he would stretch it out like a noodle along the upper edge. He repeated the mortar-lengthening process on the opposite block edge, and on the vertical edges of the adjacent block—quickly, *fft-fft*—and then, before the mortar had time to slide off, jammed the next block against it.

Next to him was a beardless youth of about sixteen, studying and imitating his teacher. Why couldn't the apprentice see how obvious the technique was? He used either too much or too little mortar, so it bulged or thinned out unevenly along his strips. The mortar fell off or his blocks wouldn't lie square. He was clumsy, slow, and overly careful. If only he would forget about trying and just plunge ahead, like his master.

My eyes at last fell on a task for a beginner like me. A man was shoveling gravel and spreading it around the floor of one of the unfinished rooms. I inquired if I could assist him.

"Sure," he said bashfully. His name was Rob. With a self-conscious smile, large bright eyes, dimples, and a red nose, he at first averted his gaze and kept his thoughts to himself. But as I spread gravel beside him awhile, he began to relax. He would occasionally rest his spade to talk. He began to try to teach me bits of Pennsylvania Dutch dialect, and as he did so, his face blossomed with pleasure as if he were discovering his own language for the first time.

Another fellow was stirring concrete alongside us, and he looked like Rob—Rob with a tanner nose and squarer features. Not surprisingly, he turned out to be Rob's brother, and he stirred the concrete as if he had three arms. Consumed by work, he uttered only the barest verbal necessaries, occasionally chuckling or raising an eyebrow as he overheard our conversation.

Rob and I finished spreading the gravel, and John moved into the hard job that came next: pouring the concrete over the floor.

Modern construction crews have long since relinquished this task to the cement truck. John, however, pushed his wheelbarrow of concrete rapidly up a skinny plank spanning the unfinished floor, and at the precise moment upturned it, releasing the molten mixture over the gravel.

The feat inspired in me a sudden bravado, and I volunteered to follow John. A tall, muscular man nearby winked at me. "I secretly wanted you to," he said. "That way I don't have to do that." I recognized him as one of Mr. Miller's sons-in-law, the one who had loaned us our furniture.

"Howie doesn't like to work," someone else chimed in sarcastically.

"Is that why he's so thin and shriveled?" I rejoined.

Howie looked at his feet and blushed. He was surely bigger and stronger than anyone here. His forearms flayed out broad and rippling, like ropes braided together. His shoulders were like mountain flanks. Yet from the little I had heard about him, I knew that he, like the Millers, had grown up surrounded by modern technology. He hadn't come from Pennsylvania, but from somewhere in Indiana, possibly near a weightlifting gym.

Another voice added, "They might say, if you work too much, you do get thin and shriveled."

"He is a little long."

"You *might* build up muscle too."

"How tall are you?" I asked him.

"Oh, about an inch over."

"Six foot one?"

"Five foot twelve and a half."

"That was a modest way of putting it."

Howie smiled wider and wider in spite of himself. Modesty was cracking.

"I remember," howled another fellow, "how ashamed I used to always get in school at weighing day, when I came out so much less than the rest. Then one day I passed Joe in length, but not in weight."

Eyes now fell on a young fellow, rail-thin, talking to us and turning the crank of the cement mixer. "I may be longer, but I *still* am skinny." The skinny one, Eli, I soon learned, was the group's scholar. Though raised in a strict Pennsylvania Amish family, his interests ranged widely beyond the community, and he kept everyone informed about world events and political developments. His voice had a certain nasal polish, a synthetic quality perfect for someone who acted as the group's radio. He asked me if Boston was still racially divided. And was "Ned Kennedy" in trouble again?

"Joe," someone said, "is another one looks like he works hard. He has a thick build. His forearms look like wood—"

"—From carryin' 'round large sacks of feed," added another.

Joe wasn't here today. But Red was. Red, at least in his own mind, may have been more deserving of renown. After I took my turn with the concrete (and barely avoided a bad spill), I came across Red. Small, spry, and dashing in looks, Red stood on a two-by-six above the new horse stalls, swaying backwards and forwards on his haunches and beckoning impatiently for more boards. I reached for a couple, but one slid from my grasp, so I asked, "Do you want me to throw up two at a time?"

From his perch, Red snickered. "You can throw up *six* at a time if you want." He uttered the taunt in a painful, pinched voice that immediately dropped to a growl. "C'mon boys," he urged, in low rumbling tones, as if trying to rev up my engine. He gathered his speech at one side of his mouth, pinched it and jeered, and then grumbled again. Next he threw his head back and flashed a brilliant set of teeth, as if to assure me he didn't really mean it. At the same time, he shifted his weight and hitched up his trousers and snorted, "Yuk, yuk, yuk, yuk, yuk." Then he resumed the low engine sounds and prodded the company once more, "C'mon boys."

He acted as though he were the foreman (which he wasn't). I could easily imagine him holding a cigar in the corner of his mouth or a pair of dice in his hand. What had he done before joining the group? Had he been a bookie? A loan shark? Someone later told me he had worked for his father on a liberal Mennonite farm. He had slopped hogs.

Until today, the Minimite men had seemed an indistinct mass to me. But now, before my eyes they were differentiating into unique characters, each with his own special role in the brotherhood. As the pieces of the building came together, so did the personalities. This was not the monolith of bearded pietism it first appeared. It was a generous slice of American diversity. But I sensed that pietistic teachings like *Gelassenheit* were not without effect. By paying close attention I realized Red was only acting the part of the punk, the person he might have been had he not deferred to the group's counsels. His swagger was too studied. *Gelassenheit* had softened his native rowdiness and given him a sense of irony to go with it. He was performing a kind of self-burlesque.

In view of the personal variation here, I wondered if the teaching of self-surrender might have advanced the cause of self in a roundabout way. By encouraging members to yield to one another, it created an atmosphere of acceptance and good humor—a place to be oneself.

It was getting hard not to let my own hair down. One fellow walked by me and grinned slyly as if we shared a secret joke. I didn't know what the joke was. His upper lip curled up when he smiled; he had bushy reddish eyebrows that met in the center and a scruffy old-man's beard on a nineteen-year-old face. When he heard someone else make a wisecrack, he wrinkled his nose, bit his tongue, and giggled. Then I found out he was Red's younger brother, Sim. Were they now, or had they ever been, in the same crap-shooting ring? Perhaps. He had red hair too, but where Red was lean, Sim was pudgy. He less swaggered than waddled. If anything (his sly look said as he padded by), he was your chum; you could *count* on him. "You're sumpin' special to us, you know?" he said to me one time later in the afternoon. I was so surprised, I didn't know how to reply.

A lanky fellow cowered in the shadow of the barn door, eyes darting here and there uncertainly. He didn't seem to have caught on. He was young, maybe eighteen, and apparently timid. I thought I smelled a greenhorn like myself.

I approached and coaxed him out of his shell. He admitted that, yes, he was new here. He had come from Alabama, where he had been raised in a typical modern environment. His father was a mechanic.

"We never stepped inside a church before coming here. Never. Well, we may have stepped in one, but not to go to a church service."

"Weren't you Christian before?"

"No. Nothing."

"What were you?"

"Go-getters. Do anything, see anything." That was his parents' philosophy. But when they learned about the Minimites a couple of years earlier, they experienced a change of heart. The community elders dissuaded his mom and dad from joining; being in their fifties, they might have been too set in their ways to make the needed adjustments. But the son was deemed eligible.

Provided he make some changes. Before moving to the area he wore his hair in the style of a punk rock musician. All his money had gone towards punk rock paraphernalia. Judas Priest was his favorite rock group. He *worshipped* Judas Priest. He had thought nothing of spending sixteen dollars on a tee-shirt with slits down the sides and a piece of cloth sewn in underneath to pass as another shirt. His white-topped tennis shoes had cost fifty-five dollars.

Today he wore black work boots, blue denim broad-fall pants that buttoned up the sides, suspenders, a wide straw hat, and a solid white shirt that buttoned at the top. A tuft of whitish frizzle on his chin tried to be a beard, and it quivered as he jerked his head like a goat, glancing at the scene around us and shifting nervous eyes from side to side. His name was Bill.

"Yeah," he said, pausing and nodding his head slowly. "It's how you stack your priorities. The Bible tells us to lead a quiet and peaceful life. Of course, you've got to do some running around. If you're going to raise a family, you've got to earn a living; there's just no getting around that."

To make his start, Bill was working as a live-in hand for the owner of this barn, the man whom I had seen laying block so skillfully. This man, middle-aged and gray-haired, was too busy to talk, but he had a story to tell. He was also a convert, and a well-educated one. He had gone to one of those expensive eastern colleges before being drafted into the Vietnam war. He was the man I had heard arguing with his wife.

• • •

Certain individuals seemed to represent a spectrum of temperaments, almost embodying the social whole in themselves. When Elbert strode onto the scene, he didn't enter it—it entered him. Everyone drifted into his orbit. He said hardly a word; his gait was smooth, legs and broad shoulders in liquid transposition, head serenely afloat.

He seemed half-oblivious to the dress code. Even though he was of strict Amish upbringing, son of the local bishop, no less, he allowed his young sons to run free without suspenders (technically required). He left one, sometimes two, of his shirt buttons open. His front tails parted and his pants drooped, revealing a bristly lower abdomen. Once I saw him pull his tails completely out, all the way up to his forehead, and mop his brow with them. Most of the men buttoned every button; he didn't seem to notice or care when he was half-naked.

Around the back corner of the barn behind a large water tank, I stumbled on a knot of men lounging around in the shade and talking near a cooler full of Kool-Aid. One of them looked up at me sheepishly and said, "We didn't yet get much done in this heat. When it's cold we like to get out there and start moving the saw to warm up. But when it's hot like this . . ."

Later in the day an elderly gentleman appeared, wearing a tattered straw hat and sandals. He was of medium height and bore himself with a slow, stately gait. His gray-streaked charcoal beard was long, his voice slow, and his eyes large, kind, and penetrating. He introduced himself as Cornelius. He told me he was single, living alone, and trying to lead a quiet life on the fringes of the community. Though not an official member, he, like Mr. Miller, felt a strong attraction towards it. I told him a little about myself, and after I had finished, he looked at me earnestly and said, "We'll have to talk again."

In addition to the adult workers, there were also the young children at play and a few older teenagers who worked alongside the grown-ups. I wasn't accustomed to seeing children in an adult work world. They gave the event an air of festivity, like cherubs adorning the corners of a picture. Except for the few older teens, they didn't directly participate; rather, they carried on a kind of sideshow with their own absorbing projects. They romped in the hay, poked at turtles beside the pond, and cavorted in and around a parked buggy, pretending they were on a cross-country trip.

Sometimes a child's shriek would interrupt the rhythm of work. Once a young girl playing near the buggy began sobbing hysterically. Quick as could be, one of the men ran and comforted her. It was nice to see that crying for a legitimate reason was not considered sinful— only fussiness during church. I presumed he was her father, but then almost any adult present could have filled the role.

As we gathered by the entranceway for the midday meal, one man with light reddish hair reached out to another young girl, pulled her up to his face, and as if forgetting there were other people present, began to cuddle her and coo. Soon a little son was on the other shoulder. They snuggled in turn as he raised them up and down his sides. When the line moved forward, he set them down again, but they tried to climb up his legs as he walked.

In the washroom three preschool-age brothers dried themselves with a single large towel as they spun inside a small corral formed by the men's legs.

Women, at this event, we knew mostly by inference. The table was set when we arrived. Their care and skill seemed to waft like an aroma from the delicacies spread upon it: the boiled new potatoes with bits of bacon; the green beans and peas; the coleslaw in heavy cream; the fresh greens, the meat and gravy; the cornbread and home-canned peaches; the variety of homemade pickles, relishes, and cheeses; the fruit and custard pies with homemade crusts. I got a plateful and began to stuff myself, but after a few swallows I felt full. The room was becoming uncomfortable; heat kills the appetite. I looked around me and saw others picking at their food as well. In a mercifully short time, grace was repeated, and we filed out-of-doors.

We had spent all of ten minutes in the ovenlike chamber; the cooks, an entire morning in it. I wondered what the wife's mood was like now.

• • •

When we returned to the barnyard, I paused to get a sense of the progress. The mason had lain a waist-high wall around the perimeter of the barn addition, allowing an opening for a door. Inside, the concrete floor was finished, smooth and drying. Outside, several six-by-six posts lay in a heap. I was puzzled by the notches that had been chiseled from the ends.

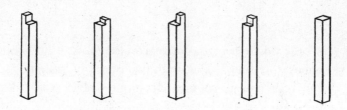

In addition, there were several other boards.

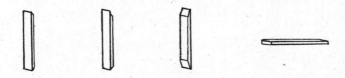

Edward, the owner of the place, began to nail together some of the notched pieces:

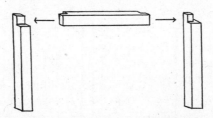

This three-piece assembly was performed twice, leaving three pieces behind:

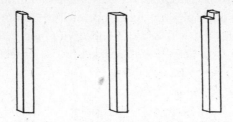

I was solicited to raise the preassembled units. The wood was heavy. I almost lost my grip, but with two others hoisting beside me, the structure rose to its proper position atop the block wall:

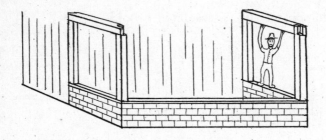

With the end-structures in place, all that remained was to insert the leftover six-by-sixes thus:

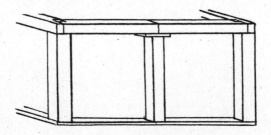

Their notched ends would interlock perfectly with the complex joints at each corner:

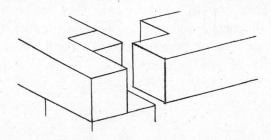

And finally, the smaller beveled two-by-fours would form diagonal braces. (Wood planks had already been bolted over the top of the block wall as a receiving surface.)

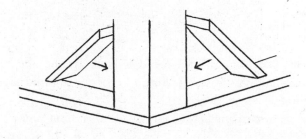

For a net effect of:

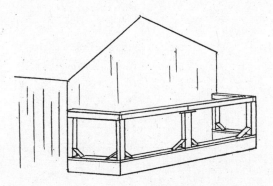

I waited in suspense to see what would happen next.

As the remaining gaps were filled in, I spent most of the afternoon handing boards up to the men overhead. They seemed to know what to do and where to go without being told. They let fly hammers on the gradually unfolding second story of the building. They pulled up the planks I passed and set them on edge to make floor joists; then laid them flat, side by side, into a spreading floor. By mid-afternoon, the picture had become:

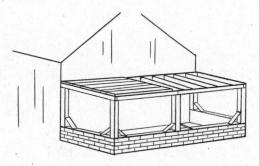

It remained only to erect a second-story frame and rafters; to unfold the building's wooden skin; and to snip tin and lay it down until it became a roof.

And from such unlikely makings, as if without anyone's trying, sprang a greater structural unity surpassing the sum of its contributors.

There even seemed to be room in the community-building for me.

the dating game

I followed Bill the neophyte one evening to a "hoeing," which is another type of work bee that tended to draw younger, unmarried people. In the moonlight, it seemed as though the corn were talking. Amid the dark rustling leaves, a soft murmur of human voices could be heard. Dimly I spied my companion wrestling with a redroot almost as tall as he was. Out of the darkness came the question: "So do you still think I'm a weakling?"

I paused to collect my wits. "Why no!" I replied. "You're forty percent muscle and sixty percent bone!" I was still trying to dispel his fears of inadequacy. Though Bill was a newcomer, he harbored hopes of spotting a prospective spouse tonight. I had come to observe that most misunderstood of human age-groups—the adolescent.

Fred, who was in earshot, asked, "What about the brain?" Fred was one of the erstwhile Amish boys.

"The brain is a muscle," Bill quickly returned.

"I didn't know that," Fred replied warily.

"Thirty percent of that forty percent!" Bill proclaimed.

"Is that why," I said, feigning puzzlement, "you have trouble standing up straight?"

Fred doubled over laughing.

I continued, "We'd better get back to the weeds. *This* is a strange one." I gave Bill's leg a gentle whack.

"Ow!" Bill shrieked.

•   •   •

The time for cookies and Kool-Aid arrived, and all the young men hovered outside the entrance to the host's house, frantically combing their hair. When the door opened, they stumbled over each other getting through it. The fairer half were already assembled inside on the far wall, conversing with distressing abandon. The males stood bashfully at the door and looked at their feet, a great gulf of untraversed floor space yawning in front of them. Bill lifted his head for a look, but in this dim kerosene light he could more imagine than see his prospects.

In sailed the cookies on a tray. A general uproar ensued as the young men rushed forward, forgetting their manners entirely. They suddenly appeared much more intent on gulping grape Kool-Aid than on getting to know their future mates. Perhaps the tension had been too much. Or perhaps they were genuinely thirsty. In all of fifteen minutes they were out the door. Had I missed something? Why had boys only talked to boys, and girls to girls?

Someone later told me the young people had gotten to know one another well enough during their long winter days in the private Minimite grammar school. These frolics merely added a little evening spice. So much for Bill's courtship hopes, however; he had attended a different school.

It was more than mildly amazing how little transpired across the gulf that separated the men from the women in that room. The boys seemed genuinely cowed by that shimmering collectivity of femininity, and the girls, for their part, were oblivious to the boys. Nor was heroic self-restraint evident among these sixteen-, seventeen-, and eighteen-year-old pubescent guys. They simply seemed too embarrassed to know what to do, like the kids at my first seventh-grade dance—too unaware of what they were feeling to know where to begin. But even in the seventh grade, skintight jeans, halter tops, and muscle shirts were the rule, and visual stimuli kept hormones at a constant state of high alert. Now, of course, children of even younger age have access to e-mail, chat rooms, pagers, and phones that make sex as easy as the click of a mouse.

But maybe there is more to life during the fleeting years of adolescence than coupling like rabbits. And as far as I could tell, the scant awareness of the facts of life did not open the gates to sexual mishap, as some might have predicted. Instead it seemed to free up youthful imaginations for other preoccupations more closely allied with their growing emotional and intellectual needs, such as the gathering together of skills and resources needed to start a family. Hence Bill's concern to raise money. One thing that struck me about the Miller teenagers, too, was how focused and self-possessed they seemed, not gangly and goofy like typical modern teens, but adultlike in everything except sexual knowledge.

The innocence of Minimite youth was no accident. It was partly the result of the care and deliberation of their parents, many of whom had relocated to this area to escape the influences that had come to predominate in other plain settlements, like Lancaster. Since children raised in Old Order Anabaptist groups do not become members, formally speaking, until and if they make the decision to join at an older age, they sometimes embark on a phase of youthful exploration called *Rumspringa*, roughly, "running around." In Lancaster, it has been reported that Amish teenage social groups have detached themselves completely from the adult world. So-called Amish gangs, with names like "Ammies," "Happy Jacks," and "Shotguns," range freely over the county.

Lancaster youth openly drive cars, wear "English" clothes, drink beer, play rock-and-roll on the radio, and join in large, raucous barn dances. John Hostetler relates one anecdote of a policeman who pulled a car over somewhere in the county. Inside were several Amish boys under the legal drinking age. After the officer checked the license of the driver, one of the passengers said to the others in Pennsylvania Dutch, "If he only knew what we had in the trunk!" The officer turned out to be of Amish upbringing himself and understood the remark. In back he found stashed several cases of beer.

Alcohol is only the mildest of the intoxicants starting to pour through some of the larger Amish districts. Several young Amish men made headlines recently after being arrested for distributing hard drugs around Lancaster.

Mr. Miller wasn't exaggerating when he admitted he had moved for the sake of his children. No longer raised on a farm or laboring under the supervision of their parents, many Lancaster youth work in the building trades and have plenty of money in their pockets—and independence. Parents look the other way when nineteen-year-old girls receive magazines like *True Confessions* in the mail or disappear for long weekends in recreational vehicles. Not surprisingly, teenage pregnancy is beginning to occur, and seems to go hand-in-hand with the freedom and privacy provided by the automobile—just as it appeared to in the wider culture, the Lynds found, when Model Ts became widespread in the 1920s.

Here among the Minimites, the practices of dating and socializing were much more structured. To prevent unwanted, unsupervised contact, not even bicycles were allowed. When boys and girls swam together, girls covered themselves ankle to neck in hand-sewn suits. Needless to say, "bundling"—the practice, accepted in some Amish districts, of sleeping together fully clad in a bed before marriage—was strictly forbidden.

To court a young woman, her suitor had first to send a letter to her parents stating his intentions. If they and the daughter approved, he was allowed to come to the house on Sunday and have dinner with the whole family. If all went well, he would be invited back a second Sunday. If he passed the test again, on later visits he was permitted time "alone" with his chosen. He could sit on the porch swing with her, or take her rowing across a pond in sight of the parlor window, or walk her up and down the lane. In all likelihood, the two already knew each other well enough from their school and church contacts, and probably from laboring get-togethers as well; or if the fellow came from outside the area, his family might well be known to hers, given the tight network of Old Order connections. In any case, young women were not obliged to accept the offers of their suitors, and sometimes didn't. But if she did agree, the marriage would follow quickly, often in two months. Parents wished to avoid any pitfalls of further delay. After being subjected to so many restrictions, the gates now swung wide open. But the couple probably had slight inkling of what lay ahead, until the informative little talks took place.

When the marriage knot was finally tied, the sexuality that had been saved for it became inseparably woven within a relationship of many strands. Judging from the large families I saw in the neighborhood, there might be a more descriptive way of putting this: the Minimite husband now freely indulged in something denied his freewheeling off-the-farm brethren—romance on the job. The low divorce rate among Minimites—zero, to be precise—may surprise us. Then again, maybe not.

TWELVE

meet the neighbors

By now it was pretty clear that the way to get to know the Minimites was to join them in work. Work of course was the currency here, and was traded back and forth like money. It was also the way to stay current, to keep in step and in good stead with the neighbors and to sometimes meet strangers.

But what a barn raising or hoeing had in spectacle, a smaller work assignment could make up for in specificity. Put differently, if at a barn raising you got to know many people superficially, in a smaller job you got to know one person well.

Edward

With this in mind I followed the trail back to Bill's home one afternoon in hopes of getting to know his boss, the man reputed to have an education, the man with the overheated wife. They lived in a symmetrical, immaculate white frame house. It almost looked like a stage prop, someone's abstract depiction of domestic bliss.

At the door I shook a hand that was more like a baseball mitt. I explained myself, and in moments I was inside. After passing through a side porch, we entered a long, bare room containing a kitchen at the near end and a living area with a rocking chair and daybed at the far end. The windows were tall and admitted much light, but the walls were blank and unadorned. The centerpiece of the spartan space was a black Pioneer Maid, which gleamed at its spot on an interior wall like the focal point of a simple chapel.

With his hat off, Edward's ample gray hair looked like the mane of an elderly lion. He stood an inch or so taller than I and had square, slightly hunching shoulders. His eyes were friendly, yet they glinted with the slightest hint of . . . what was it . . . mischief? Was it that he already knew who I was—maybe more about me than I wanted him to? Or was there something about his own past he wanted to divulge?

A woman in a white head covering stepped from the kitchen area. Two wide-eyed young things clung to her apron, each wearing a miniature version of the same headpiece. Mr. Pendleton introduced the woman as his wife, Grace.

She didn't look angry at all. If anything, she was quick to catch my eyes with the pleasure she took in my arrival. Roundly proportioned, with a round untroubled face and dimpled cheeks, she could have sat in for the portraiture of the Mona Lisa, were it not for the head covering.

As Grace prepared mint tea, Edward and I seated ourselves on the daybed, and after I told him a little about myself, he filled me in on his background:

"I grew up in a good Catholic family in a semirural area," he said. "My dad believed in hard work, and he even found a local farmer, a Jew, whom I could work for in the summer. My mother, bless her soul, was a gem of a person. She's dead now, but my father put her on a pedestal.

"When I went to college, I came with all these ideals; I started out loyal to the Church's teachings. But it was the sixties, and that's when things began to unravel. I would be one person at home and another altogether at school. My parents had no idea what was going on. I did things there that I don't even want to think about.

"After I graduated, I went to Vietnam. I was a combat engineer, so I didn't have to do any of the fighting myself, but that didn't mean things were hunky-dory. It was war; we were soldiers. You could get a beer for free. Soda pop was five cents. Drugs were plentiful. The master sergeant was a pusher. The government supported that."

"Wouldn't that make the soldiers less alert?"

"The government recognized that they needed a way of blotting out the horribleness of the experience." He had since learned, he said, that sixty percent of Vietnam veterans who needed help after the war had returned with drug and alcohol addictions. Edward didn't get into hard drugs, but he still had much to forget.

"It was dehumanizing. Take what they used for a"—he pointed through the window and I saw the outhouse—"and called something else."

"You mean the latrine?"

"Yes. But they called it something else."

"Outhouse? Backhouse? Shed?"

"They named it for what went into it."

I snickered.

"They took a long board and cut ten holes in it. Then they took those fifty-five-gallon drums and cut them in half, and put them below."

"And that was it? No dividers in between?"

"No. You just went in there and sat next to whoever else was there. And it was so hot, right under the sun."

"It was outside?"

"Yes. It had a roof over it. When I first got there, I'd go way, way early in the morning or way, way late at night. But after a while I got as sloppy as the rest of them. I'd march up right after supper at the hottest part of the day. And there I'd sit with maybe ten people beside me reading the newspaper or conversing. And the stench!

"There was a fellow whose name sounded like what they called the latrine"—he contorted his face and tried to speak something in a mangled Chinese accent—"and when those fifty-five-gallon drums got full, he'd take them away and burn it. Oh, that was the best time to go. He'd burn it not twenty feet from us, and the smoke would just come rolling in. Then you could enjoy both the smell of what we were doing and the smell of what he was burning. It was so dehumanizing."

"The word 'human,' " I corrected him, "comes from the same root as the word 'humus.' It was humanizing."

Edward paused, barely smirking. "It was humanizing.

"And the shower. There was absolutely no privacy. They ran a pipe through the center and water would just come straight down from a 'T.' It would just come in a stream on you. There was just the shower and two dressing rooms. And they'd have women come in to do our laundry. And they'd do it in the dressing room because there was water there. There again, at first I'd only go early in the morning or late in the evening."

"Could they see you?"

"You were right *there*."

"I thought you said they were in the dressing room."

"They'd be sitting right here, and you'd be taking a shower right there, hearing 'zchulbtha.'" He held his hands a yard apart. "They needed the water for their laundry. Every now and then they'd reach over and get some."

"At first we'd say nasty things or point the shower nozzle at them. But I"—his Adam's apple bobbed—"got used to that too."

Edward remained impassive as he spoke, as if the trauma of it all went without saying. But I wasn't sure I understood . . . While his buddies were dying in the jungle, he had shared a shower room with Vietnamese laundry women?

"My last year of duty, I took a leave in Hong Kong. I went to a tailor and had two suits made. One was blue and one was black— business-type suits. Then, when it was finally time to fly back to the U.S., well, we had to wear our *uniforms* to get into the airport. So I brought the suits in a carry-on bag.

"Then if we wanted to get on the plane, we had to wear the uniform. But once I got on, I left my seat, took one of those suits, and changed in the restroom. A lady was sitting next to me. She didn't appreciate it much, I could tell.

"My parents were at the airport to pick me up. Of course they were expecting the works—me laden down with medals." He sank in his chair to show what he meant. "As they stood there, the first-class passengers went by. Then so did everybody else. And I wasn't there. Of course they were looking for a soldier in full regalia. They were a little disappointed.

"The war, you see, made me lose faith in all institutions. When I got home, my mother asked me if I was going to wear my uniform to church. I had a hard time responding because I wasn't planning on going to church."

One thing led to another, he explained, and he gradually began moving in Mennonite circles. Then he met Grace, who though raised Mennonite had made an unusual step into college. After marrying, their search for an authentic Christian community finally landed them here. There were several attractions for them. First and foremost was the group's adherence to "non-resistance"—the refusal to bear arms of any kind or even to defend oneself from an aggressor. Second, of course, was an interest similar to my own, the elemental, untechnological way of life, the close ties of a face-to-face community. The third attraction, however, left me uneasy. They honestly agreed with the group's interpretations of Scripture and its refutations of Catholic positions. Unlike Mr. Miller, Edward had no problem accepting this small local assembly as the only true Christian church, whose doctrinal and disciplinary proclamations could not be gainsaid. He then provided me a scriptural basis for this belief, which I had a hard time following.

It was not easy for me to understand how someone with his intellectual aptitude could have accepted this. The only way I could make sense of it was that he had become an enthusiast. He had, somewhere on the bumpy ride from the battlefields, crossed a subtle line—not only to live the way of life, not only to love it, but to *idolize* it. All too ironic for a lapsed Catholic. But with so much in common between us, I could sense an unspoken bond already forming. I knew I was extreme in my own way and had come well nigh as far as he had. I sympathized.

When he was done with his story, he led me back out to the barnyard. There I gazed upon more evidence of his single-mindedness: fresh paint on the buildings, well-swept stalls and shop, orderly arrangement of tools. At an outdoor storage closet he paused and extracted two hoes; he mentioned he had some catching up to do and gestured towards the cornfield. "How would you like to—?"

My request to join him in work was being granted. But I sensed more than this: that Edward was proposing a sort of test of my mettle. I hesitated, wondering where this was leading. Could I possibly meet his rigorous work standards? I felt a tingle of adrenaline.

We strode over to the field and took aim. What ensued was like a hockey match. The challenge was all-absorbing, mentally as well as physically. Every weed posed its own size, shape, thorniness, and root structure, as well as distance from the corn. Each called forth a fresh judgment for the mind, a new nimble jab of the hoe. We finished a row; the sunlight was ebbing. We circled back, neck and neck. I batted corn leaves from my eyes and kept my line of vision clear. I gasped and slashed. If I could just hold out a bit farther.

There! Edward finished only slightly ahead. It had been a good showing. My heart was pumping, my lungs heaving. In the afterglow of success, I realized where I had seen a face like his before—the set jaw, the wry grin. Here was the visage of my college rowing coach, which of course made me question his pacifism.

A little later Bill joined us. The weeding was light but the field was huge, so there was plenty of work for three. We stayed together in a clump. To keep up with Edward, half-consciously I had refined my technique. I alternated between hoeing and bending over to pull by hand. Whenever I bent, or sometimes kneeled, to pull a large weed, I would throw the hoe on the path in front of me, grabbing it again quickly when I stood up.

With enough practice it was possible to talk even at this pace, though not easily. As I worked I strained to hear bits of what Bill was saying. My ears perked up when he began to mention something about the latest, most asinine application of modern technology.

"There was a woman in Missouri who just finished washing her dog," I heard Bill say. "It was cold out, so she thought she'd stick him in the microwave to dry. She wanted to stick him in a moment and let him out. So she put him in. Shut the door, and opened it right away. But he had cracked open."

"I can't believe that," I cut in. "It would have taken longer."

"No, it really happened. I read it in the newspaper."

"It couldn't cook you that fast."

"She had it turned on full force."

"But if I stick a hot dog in for just one second, it's just as cold when I take it out as when I put it in."

"Maybe it was a minute."

By now we had put our hoes down and paused, and in the lull of activity Edward's voice boomed from a few rows away: "But that was a hot dog. We're talking about a cold dog."

Edward had made a pun? I turned to look. I saw him between the stalks, covering his mouth in embarrassment as if he had just burped. "There's a difference," he added half-defensively.

"What a headline that must have made," Bill clucked. "HOT DOG FOR DINNER."

After the conversation, Edward approached me and apologized. He explained that there was a scripture against jokes; it sorely grieved him that he had weakened in my presence, and he hoped he hadn't set a bad example. For goodness sake . . .

As if to make up for his lapse, the mood on the farm the next time I visited was much more somber. I found Bill alone in a vast field, shocking oats. This consisted of collecting bundles of oat stalks (already cut with a horse-drawn device called a binder) and setting them in piles scattered throughout the field. Each pile, or shock, comprised a number of bundles set vertically, capped by two laid sideways to shed rainwater. The work gradually became rhythmic and comfortably automatic, but for once I couldn't relish the conversation.

"I'm not like the others here," Bill grumbled. "The other fellahs, when they start out, they get a lot of help. Eli Fisher, when his sons turn twenty-one, or when they are about to get married—well, they can get it sooner if they need it—he gives them each six hundred dollars. Our parents are too poor to give us anything. Everyone else here is way ahead of me. When I got here, I was starting from scratch."

To make matters worse, he explained, Edward gave him terms less favorable than those of other *hired* hands. Another young man he knew, also a novice, worked in a family that provided room and board in return for one full day of work, leaving him the remaining five workdays to earn money on his own projects. Bill had to work three full days for Edward to earn his keep. And these were not, in Edward's case, neatly bounded eight-hour workdays. They could easily last ten or twelve hours. (In addition to the land Edward owned, he rented several adjacent fields besides.)

As if to prove Bill's point, the shocking we were doing now (on a rented field) had to be finished today however long it took. It had already been near suppertime when I joined him. "How much more?" I asked.

Bill's eyes narrowed. "There's two more fields, and one is bigger than all the others combined. It's depressing just to think about it. I'd almost rather be in Siberia."

There was a long pause.

"But then," Bill said, continuing his thought, "there's a prison there."

"That's just what I was going to ask. Is even prison better, or is it the cold you want?"

"After doing this awhile, a cool dark cell might seem like a vacation."

As daylight ebbed, Bill began dreaming of ways to get even. He had been reading a history of the American Indians and mentioned two methods they used of enforcing criminal justice. One was the squaws' gauntlet, and the other was the braves' gauntlet. The squaws only clubbed their victims. The braves—. Bill caught himself and said, "Well, never mind."

Remembering something I'd learned in graduate school, I told him why historians thought native America had become so warlike. The accidental introduction of horses by Coronado had drastically expanded the hunting range of every tribe and intensified the competition for land. When guns were brought in later, the effect was like a spark in a powder keg. In combination, the two artificial intrusions amounted to a textbook case of wholesale cultural change via almost "pure" technological causation and a breathtaking rebuttal of the claim often made that technology is a "neutral" agent of human intentions.

When I finished talking, Bill, between huffs, replied, "I knew that already." He was huffing because we were busy running through our own gauntlet of oat sheaves in the attempt to finish while we could still see.

We didn't make it. The last two rows of shocks we did in the dark. They might as well have been put up by blind men.

•   •   •

In the few free hours a week that were allotted him, Bill scrambled to get a tomato patch going. Tomatoes, if timed well to demand, could be quite profitable.

At last they began to ripen. I wanted to give him all the help I could, so I joined him and Grace picking. With no small exhilaration we burrowed into the vines, filling bucket upon bucket. A steady breeze cooled us. Tall rows of plants acted as dividers channeling the air in our direction.

"Out in the fresh air, seeing the fruits of our labors," exclaimed Grace. "Before it gets warm," she added.

"I got three buckets just in the first third of a row," Bill gushed. "Whew!"

"Grace and I got seven in our first third!" I returned. "How much they payin' you a bucket?"

"Ten dollars."

"A rich man in our midst," boomed a deeper voice—Edward's—as he happened by.

The tomatoes broke from the plants with a delicious crunch. So plump were they, many had become inextricably trapped between stake and vine, or vine and branch, and exploded when we tried to loosen them.

It took an entire box wagon to haul them over to the community produce distributor.

After lunch the distributor stopped in with some news. He looked grim. Bill's tomatoes were overripe, he explained. None of the buyers wanted them.

What Bill didn't know was that wholesale tomatoes have to be picked at the pink stage. He had waited until they were red. But by the time his patch had produced enough pink ones to make a load again, the price had dropped by half. I wasn't sure why Edward hadn't bothered to inform Bill of any of this, but I did know Ed was a firm believer in paying for one's own mistakes.

I next found Grace in the barnyard bearing an armload of leftover ripe tomatoes.

"You know what Bill's been doing with these?" she cried, creases deepening from the corners of her mouth.

"Throwing them over the fence?"

"No! Squashing them with his feet."

When Edward marched by, she had to act fast. "I need you to help with the tomatoes," she said brusquely. There was a hardness to her voice implying an "or else." Clearly she had gone through this routine before.

"O woe!" Edward moaned, wincing. He couldn't let her off too easy, so he hammed up the slight to his dignity. Soon he was beside me gathering Bill's rejected tomatoes from a table in the wood shop. Their destiny was the big kettle in the wash house. Edward began to wash them. But he was taking them one at a time, scrubbing them like potatoes. Huh? Hadn't he seen Grace just now dump a dozen in a bowl of water, give it a little slosh, and empty it again?

Next came slicing. I did what Grace had showed me: a few quick slashes and into the next bowl.

"Aren't you going to cut out the scars?" Edward asked.

"Grace told me it didn't matter."

"Grace!" he wailed over to the next room. "Don't you want us to cut out the scars?"

"No," came the voice. "They come out through the grinder with the pulp."

"That's not fair," he whimpered. "You told Eric an easier way."

In his forties and as ignorant as Bill about tomatoes? I got it. He was playing dumb!

But the strategy backfired. I wasn't really listening to what was said next until Grace's voice began to pierce the air like a needle: "I don't understand where you come from saying it's my responsibility to take care of the whole garden without any help from anyone else. Name one other woman in this community who has to do that." I turned to look at her and as quickly turned away. Her face had gone white and the veins stood out near her temples.

Edward's eyes were half-closed. Finally he said, "Do you want them alphabetically or how?"

Grace volleyed a lightning-fast comeback; Edward rejoined. Fireworks . . .

After Grace left the room to cry, Edward turned to me and took in a deep breath. "Grace needs all the encouragement we can give her. Not the kind I've been giving." But he added in a soft voice, in partial self-exoneration, "It's her arthritis."

As if reading my thoughts, he blathered on: "Grace makes fun of me because when I start working, I can really get at it. Then at the end of the day, I just lie there. I just *really* enjoy that. I can read the newspaper or a book. But Grace's still working until nine or ten at night, puttering around. And then the next morning, she says she's so tired. I try to tell her to manage better, and she'd get it done faster. But if you saw her family. When they get together . . ." He slowly opened and closed his hand to demonstrate a yapping motion, and rolled his eyes. "I'm not saying it's bad. It's just the way they are. They like to talk, and the work just comes afterwards."

The more I got to know Edward and Grace, the harder it became to resist this conclusion: they were an example of how not to get involved in a Minimite community. Edward's tendency was to extremes—from breakneck work to self-satisfied indolence, from impish wit to deadpan sobriety, and from religious disillusionment to iron absolutism. Grace, for her part, was a lovely soul, but she couldn't keep up the pace. She kept getting caught in the spokes of his shifting cycles.

They didn't so much need community; they needed counseling.

Harvey

One day there came a knock at the door—a heavier and slower knock than I was used to hearing. In the doorway stood a young man of burly bearing, smiling widely. He introduced himself as Harvey, Mr. Miller's oldest, married son. And he had come to sell me a pig.

"A pig?" I asked.

"Don't you have a lot of extra milk you're just letting go to waste?" he returned. "That's what I heard." (Waste was anathema to Minimites of all stripes.) Harvey pointed out that if we started feeding a small pig now, we might have our own homegrown meat supply before Thanksgiving. He could sell me the shoat cheap because it had a rupture and wouldn't last much longer in a pen with other, prodding pigs. Twenty dollars for the pig.

I told him I'd think about it.

I hated wasting milk, and I loved the thought of a fresh meat supply. My only hesitation was pecuniary. Twenty dollars was forty percent of our monthly food budget.

After considering the proposition a few days, I met Harvey again in Mr. Miller's barnyard. This time his wife and child were with him on a spring wagon. I walked up to them and said hi. It wasn't hard to get a conversation going. When he chuckled, two large dimples formed on either side of his mouth and his well-rounded belly shook noticeably. He was not nearly as reserved as his dad; in fact, he was downright jolly. I reminded him we hadn't completed our pig transaction yet, but if he pleased, I preferred to pay in labor. He didn't think but a minute. Suddenly I was being whisked away on his wagon.

We passed his father on the way out, pausing long enough for Mr. Miller to warn me: "Watch out for that guy. Make sure they give you enough to eat. He's liable to eat the food off your plate."

Harvey chuckled, and off we drove.

I mentioned, swaying as we rounded a bend, that he was lucky to have such an insightful father.

"You're right." Harvey sighed good-naturedly. "When I was younger I rebelled, but now that I'm away I appreciate him. From a distance!" He chuckled and heehawed. "No, I really should consider myself fortunate. Could've done a lot worse."

Harvey and Gertie, who appeared to be in their late twenties, occupied the wagon seat with their baby while I perched on an overturned bucket behind them. For a while it was all downhill on a narrow gravel road, and the horse was not timid. I struggled to keep my balance as we lurched around curves and over bumps. We forded creeks and veered near the edges of concrete overpasses, and at one tight corner I almost thought we'd plunge into a ravine.

They lived some distance away from the established community, so there was time for many more quips and heehaws. Affable, relaxed in his manhood, Harvey raised important questions for geneticists: how could someone with his dad's personal and physical traits have sired him?

If her husband was large and ample of girth, Gertie was petite and angular in shape. Even her face was small, with sharp features. But when she opened her mouth to laugh, she was as raucous as her husband.

The two of them were goofy and proud of it. No subject of conversation was too madcap. After inquiring about my family back home, they zeroed in on the fact that I had changed my younger siblings' diapers.

"I do that too," Harvey said.

"Right now, with him?" I asked, indicating the blond child sleeping blissfully in Gertie's arms, like some baby sun-god of the north.

"Yes." Then he added, "But if it's the messy kind, I give it to her." He motioned Gertie's way, and she giggled.

"You must have to clean up worse messes than that! Don't you have some pigs or cattle?"

"Yes, but human is worse—to me."

We finally came in sight of a small white bungalow centered on about forty fecund acres. The house was almost indistinguishable in size and appearance from the one I was renting from Mr. Miller. Would all the Miller offspring begin their marriages in the same adorable starter cottages? Inside there were differences: the kitchen was in front and the bedroom in back. The walls were painted white, not paneled in dark wood.

Harvey and I settled into our chairs at the table while Gertie began laying out dishes and utensils. "We want you to be able to say you got enough," she said. I noticed some nervousness in her voice. In a few minutes she had set before us bowls of homemade cottage cheese, cabbage and carrot soup, chili beans and meat, diced potatoes, sweet pickles, okra pickles, pickled corn, and fresh-baked bread and butter with a jar of homegrown sorghum molasses. I took a bit of everything except the okra pickles, pickled corn, and molasses. It wasn't that I didn't care for these; I just ran out of room on my plate.

Partway through the meal Gertie went to the counter and returned with a serving dish of diced cucumbers, which she placed in front of me. She looked at her husband and turned up her nose. "You don't like raw cucumbers."

"I don't ask for them!" he squawked genially.

She seemed to be tweaking him, perhaps because of his obvious preference for heavier foods. He was on his third slice of her home-made bread, with home-churned butter and homemade elderberry jam on top. "I don't like boughten bread," he chortled. "Why, I wouldn't even eat bread if it was boughten." And he kept heaping the beans and more bread onto his plate. Compared to the other Minimites, he admitted, the two of them did eat noticeably more. "But the best way to lose weight is to work hard!" he asserted.

"I don't know!" squealed Gertie, "Your pants are sometimes pretty tight!"

Church was a penance for Harvey, he confessed, because it meant he had to go four straight hours without eating. He could not make it through the service without severe hunger pains. "It takes seven days to feed me for six days of work," he said. "How about you? Getting to like farming?"

"Pretty well. Gives you a little of everything, including being part of the family. I like not being completely separated from my wife. That's probably why my parents divorced. Of course that's not all the reason."

"But it's a start," said Harvey.

"I like having Harvey around," added Gertie affectionately. "And he knows it." Then she arched her eyebrows. "Otherwise I wouldn't know what he was doing."

Harvey grabbed his son from the high chair and began cuddling him unabashedly. The tiny boy, in turn, rejoiced in the attention. Harvey threw him up and down on his lap and nuzzled him with his beard. I was won over by the exhibitionism of it.

The boy, Joshua, had started walking two weeks earlier but had regressed since. "One thing about children," Harvey offered, "they learn things only one step at a time, and they will never learn something before they're ready. There's no use in trying."

For the finale, Gertie brought out a selection of desserts: persimmon pudding, a mixture of strawberries and peaches, and homemade chocolate chip cookies. I think I ate as much of this course as Harvey did. It was gratifying to partake at the table of a healthy marriage.

●　　●　　●

After the meal, I was ready for the labor Harvey had promised. I followed him outside to a storage shed, where he grabbed a hatchet and a bow saw. Harvey, I sensed, was brighter than he let on. His clunky good nature was deceptive. Listening closely, one could hear him juxtapose words like "pig barn" with "whereas"; "he don't" with "materialize." Other words he slipped in included "conception," "disregard," "advocate," and "acquire."

"I have to confess," he admitted. "I am a bookworm." All through school he pretended to read his assigned texts but concealed pleasure books behind them. Biography, nature, and some fiction were his favorite subjects. "I could sit all day with a book, if it's something I like." He explained that he used slang expressions when talking with the locals but upgraded his vocabulary when addressing better-educated people.

We were meandering our way out of the bright midday light down an incline into a Hansel and Gretel wood, quiet and dark and green, with thick cedar trunks beginning to appear in the undergrowth like macabre wooden legs. Crickets chirped softly and occasionally a lone blue jay cackled in the stillness. One gnarled tree by an old tumbledown fence gave out an enormous show of boughs like a witch's broom turned upside down. After picking our way through scratchy underbrush and murky bogs, we came to a tall, straight cedar. This would make a good pole for the pig barn Harvey planned to build.

He let me watch at first. He took his axe, and in what looked like some sort of karate demonstration, whacked furiously until all the lower branches were chopped into smithereens. Then he set the blade to the side of the trunk and reared back. *Chop! Chop!* A deep cavity appeared in the side of the tree. He next picked up the bow saw and directed me to hold one end. We placed the blade against the trunk opposite the cavity, and the saw began to move. I couldn't reciprocate as smoothly as I would have liked. Harvey pushed with a confident ease, but my returns were unsteady and irregular, and the blade frequently seized. Still, since the trunk was not thick, the tree soon began to yaw. With a tremendous "swoooooossSSSHHH!" it crashed through the brush and landed on the ground. A captivating stillness followed.

I took a turn with the axe on the next tree. The first swing went wide of the mark. The second failed to separate the limb. I looked up at Harvey importuningly. "I am a little rough," I confessed.

"It's not something you should be ashamed of," he assured me.

After a few more misses it became apparent we would make much faster progress if I left the hatcheting to him. Despite the cool and darkness and the "help" I was giving him, during the sixth round of hatcheting a stain began to spread from around his suspenders, darkening his blue shirt. "I'm getting wet," he announced.

"Do you wish you had a chain saw?"

"No. Used to have one. Didn't like it." A chain saw would have been as tiring, he explained, but in a different way. It wasn't the physical exertion so much as the stress from the noise and vibration. After chainsawing for hours at a time when he had lived with his parents in their former progressive Mennonite settlement, he felt "numb all over. And"—he pointed to the bow saw—"when I stop, it stops too."

There remained a larger question I'd been meaning to ask, one principal puzzle of his beautiful, quiet, and solitary farm. Harvey lived two miles from the nearest Minimite neighbor. Everyone else in the community lived within a stone's throw of each other. For someone as gregarious as he, wasn't this a little odd? As we began to lug the stripped poles back towards the house, I asked, "Why do you live so far away from everyone else?"

It was a matter of pragmatics, he told me. When he was looking for a farm to buy, he narrowed his choices down to two. The other lay in the heart of the settlement, but it was twice as much per acre. The difference, more precisely, came to $30,000 for forty acres versus $38,000 for seventy-one acres. He was able to keep the larger tract in tillage for the time being by going into partnership with his brother; then at some later date he would divide the farm with his son (the baby he'd been nuzzling). He chuckled. "Since I have it, I wanna keep it; but if I didn't have it, I wouldn't miss it."

I heard a loud rumble from Harvey's direction. He smiled at me knowingly and said, "Remember the bumper sticker they had during the energy crisis? 'Conserve natural gas—fart.'"

I gave the due grimace. The alternation between shrewd reasonings and country crudities came a little fast. I resumed the thread.

"But being this far away," I asked, "don't you find it hard to get people to help you?" Surely it would deter trading favors to have to travel several miles each time it happened.

"Yes, but look at it the other way. Other people don't ask me to help as often. And you don't get involved in some of those things that go on."

What did he mean by "some of those things"? Minor frictions, differences of perception of value in traded labor, perhaps? Or was he referring to deeper squabbles over fundamental issues? I sensed that he, like his father, had a more easygoing attitude about doctrine than some of the other members here and would just as soon sidestep debate. I also noticed that his farm was located at the opposite end of the community from Edward's. That explained his absence at the barn raising.

Whatever the answers, the differences between himself and his father were smaller than they appeared. Like his dad, he sought sensible solutions to practical problems, studying how to maximize benefits while minimizing trouble. He also had some level of genuine faith that directed him in his decisions on behalf of himself and his family, and he willingly endured the necessary ritual on Sunday. But he was also a person who would gladly take a day off from work to read a good book, and not lose a wink of sleep over labors lost.

As for his preference for the bigger, cheaper piece of land, this turned out to be a passing phase. In days to come he traded his sizable farm for a smaller tract near his father's homestead, thus giving up his serene isolation for the convenience of closer neighbors. One of whom was me. At the end of the day I was one pig richer, and Harvey had gotten the help—and company—he needed.

Cornelius

The weather was unseasonably cool and bleak the day I set out for the house. I walked along back roads edged by walls of corn so tall it prevented all but an occasional glimpse of the surrounding landscape. I shivered in the cool breeze, wishing I had brought a sweater. The weather reminded me how important a wood supply would be this winter. I was glad I remembered to bring a dull saw.

I finally came upon a ramshackle hut in a clearing. Half of the front porch was gone. A corner post hung unsupported from the roof. Spindly trees were growing up around the tilting foundation; plastic covered the windows instead of glass. I was about to pass by, but then I noticed a tin plate, fork, knife, spatula, and black iron skillet neatly arranged on the back porch floor, apparently laid out to dry. Then I saw some movement.

It was Cornelius, with his long gray and charcoal beard. I recognized him from the barn raising. He had come out the back door.

He looked up in surprise, then brightened when he saw who it was. "I would have come to see you," he said, "if you hadn't come first."

I joined him on his porch, and after we shook hands, he signaled me to the door. The interior of the cabin was unexpectedly inviting: dark-stained walls (quite solid-looking and plumb), a desk heaped with old books, a comfortable sofa, armchairs, and a period-piece black woodstove. It was as though the facade of the building had been merely a ruse calculated to ward off all but the most determined ascetics. As I sat on the sofa, he took a seat by the kitchen table. He saw my teeth were chattering and within minutes got a good fire roaring in the stove.

We exchanged a few pleasantries. He seemed to breathe his words rather than speak them, in a voice deep and quavering. Then, with little ado he went to the point, as if he were resuming a conversation we had recently left off.

The Minimite community, he bemoaned, was too affluent! "In the Bible you don't find any reference to the use of horses. They were used then, but not by the early Christians. Christ rode on a donkey one time, but they had to go find someone to borrow it from. That shows they didn't own any themselves. The only way they traveled was on foot or by ship. *On foot or by ship.*" The last words he repeated didactically, as if trying to impress the lesson upon me.

He realized that life had changed after two thousand years, but he felt that the standard of living in biblical times was adequate. No one in the local community, he admitted, paid much attention to him.

"But there's only one of you in this house," I said, half-jesting, "Maybe if there were twelve they'd listen more closely."

He nodded and laughed. Then he explained how he had gradually arrived at his present viewpoint. He too had grown up in the Lancaster settlement, which he viewed now as egregiously worldly, a mecca of hedonism. There he had had a good-paying job. At that time, even though he was surrounded by farmland, he bought all his food, at the grocery store. He'd always purchase whatever tempted him on the spur of the moment, telling himself he had a good income, he could afford it, he could take life easy. But he gradually noticed the "reverse" taking place. Though his job gave him flexible hours, he inevitably worked overtime in order to buy more refined, processed foods.

"I've noticed," he said, "that's generally the way it goes. Most of the work around here is just to turn the machine. *To turn the machine.*" He repeated the phrase with a wistful gravity. Here he had come to the crux of his position. On his own, living among the prosperous descendants of Menno, with no vast knowledge of sociology, he'd arrived at an observation not unlike that of Max Weber in his famous study *The Protestant Ethic and the Spirit of Capitalism*: the means by which the spiritually reborn demonstrated their righteousness were threatening to become their ends. The outward tokens of salvation or self-worth were starting to stand in place of the inner reality.

I almost murmured something about Edward. But even Cornelius admitted to having once succumbed. It was only a chance event that had led him to break the pattern. His parents decided to move to another county in Pennsylvania, and it was there, in his new position as schoolteacher, that he had reason to reread the Bible. One day he read the passage from Acts that told how the early Christians gave away their possessions and commenced to "hold all things in common." Cornelius began to rethink his priorities and to scale back on his consumption.

His quest for simplicity finally led him here. Now he owned practically nothing and bartered for the things he did have. In return for teaching school for free and contributing to occasional work bees, he received a place to live, space for a garden, and use of the adjoining timber lot. He ate no meat or butter because his arteries were hardening and his pulmonary vessel was "partly blocked up." This, alas, had compelled him to consider giving up teaching. But the neighbors would continue to help him out.

His life was as simple and serene as Thoreau's, it seemed, with the difference that he acknowledged its dependence on community. (Thoreau accepted hand-outs too but, alas, kept quiet about their source—his friends and relatives in nearby Concord!)

"Do you think you're getting enough protein?" I asked.

He thought so. He took care to include a mix of homegrown corn and kidney beans in his diet, in addition to the milk.

Although it wasn't perfect, he still held the local community in high regard. At least it was better than any other he knew of in the United States. Its ban on owning motors made a huge difference. The Lancaster Amish, he said, could not imagine anyone surviving without motors. "They're just having a hard time making do as it is!"

I agreed that going motorless was critical. To me the distinction between the hand tool and the self-operating machine explained a lot about the unraveling of modern society. I suggested to Cornelius that, for the Lancastrians, the hidden costs of automated devices went hand in hand with an upward spiral of material wants.

"Yes. But they just don't see it." He paused thoughtfully. "You couldn't do what we do without Christianity."

I hesitated. Kalahari bushmen came to mind. I told Cornelius what I had learned about the !Kung tribe while in school: that by gathering nuts and hunting, they were able to meet all their material wants while working only two or three hours a day. "If we are going in circles," I said, "if we really thought about it, we would realize how silly it is. Simplicity is simply the intelligent or wise thing to do."

He thought a moment and nodded. "I suppose you're right. The point I was trying to make was that if you did it, it would have no values without Christianity. There are hippies who live much like we do, but they do all kind of other things." He cocked his head a minute, reconsidering. "If we were really Christian, it would result in this."

He shuddered at the moral condition of the present world. "One time I was in a hospital under observation for a week. I was in a ward with ten beds, and up in the corner there was a television set. They'd come in and turn it on at noon, and that thing would stay on 'til midnight. I'd watch for a few hours, then finally get disgusted and turn over in my bed. But I noticed that if I watched some programs, it got 'til I waited 'til the same time the next day to see how it would continue." He let out a humorous "humph." "I remember one that was called the *Million Dollar Movie*."

"You mean the *Six Million Dollar Man*?"

"No. Believe it was the other. It had a big-time evangelist and all these good-looking girls, and he would go around killing them at night. He spent half his time preaching in front of the congregation and the other half involved in mob dealings. He finally even killed all of them that were in on it with him, and he was driving real fast down to a boat to make his getaway." He stopped short, as if he realized that what followed was too horrible to relate.

But I couldn't sit still. Finally I asked, "Did he get away?" Having prided myself on being weaned from television, I was slightly embarrassed to have to beg for the ending.

"There was a police blockade there, and when he came through"—he lowered his voice to a growl—"they just gunned him down."

We both fell silent for a moment, as if pondering in this grisly example the death of America's claim to moral leadership.

"There was another one with a real tall man and his father, and they were into witchcraft. They had an extra-good-looking daughter, and they were trying to find a husband for her. One day they found a young man for her, and they got him onto the porch. He took one look at the taller man and jumped right over the rail and got right away."

"What made him do that?"

"It was just the way he looked."

I realized he must have been talking about *The Munsters*.

"Then these two men decided they'd create a husband for her. So they went down in the cellar, and they took a frog and put it in a dish. Sometime later this repairman came to check on something, and the little boy took him down to the cellar. Then the two men came downstairs and saw him there. They thought he was the husband they made for their daughter"—his voice cracked and he emitted a jovial squeak—"and they caught him and tied him up."

Television clearly had caught Cornelius's fancy; but at least he had an excuse. His captivity had been unwilling; that of most Americans was voluntary.

Still, I found it remarkable that his recollection of these two TV episodes remained so vivid after many years—it was as if they were scenes from his own past. It was also striking that his few memories of the technological world included one of Frankenstein.

He took me outside and showed me his blade-sharpening apparatus. It consisted of a vise mounted on a stand, and a file. He demonstrated how to sharpen a saw, then gave me a turn. I ran the file at forty-five degrees in the crevice between every tooth, alternating directions depending on whether the tooth pointed to the left or to the right. Conversation flowed freely as I played with the file and ran it through my fingers. And the action of sharpening added a subtext to Cornelius's message: the community might have even *more* leisure if it just scaled technology back a *bit* further.

After getting to know only three of my neighbors better, I was becoming more certain of an impression I had already received at the barn raising and the beard-bragging session: that there was more than one way to live with less technology. The common ground held by the different members of this community, indeed, was still being mapped out. As things stood now, what joined them together, beyond certain basic positions on Christianity and the use of machinery, was a willingness to differ.

Edna

Time wasn't diminishing Mary's gusto. Far from wearing her down, a more physical life had notably improved her health. After a month or two on the farm, a mysterious stomach ailment she had suffered in Boston disappeared on its own. Whether it was the fresh air, the exercise, the slower pace, the organic vegetables, or a combination of the above, we didn't know. What we did know was that she had never felt better.

In this flowering of health and hardiness, pregnancy was affecting her little. After that initial twinge of morning sickness, she cured the problem by eating more. We had read that the cause of this malady is not food but its lack, as the body's metabolism changes to feed two persons instead of one. She simply ate the nausea away.

Her main lingering symptom was external: she showed early and large. Though a hardy one, Mary is slender and small-boned. There was nowhere for the baby to go but forward. That new protuberance altered the physics of her posture, and the one real problem pregnancy posed was back strain.

Needless to say, as month yielded to month, this problem didn't go away but worsened. It became harder and harder for Mary to bend in the garden, and finally even to walk. Bicycling was an antidote; the steady keel of the wheels reduced bouncing. But since she was carrying more weight, she couldn't keep up with me when we biked together, so I located a used tandem, and voilà: we soon found we loved getting around on the bicycle-built-for-two. On Sundays we would noodle all over the community, making visits. We discovered one long round-trip loop from our house that seemed to go downhill the whole way. (There was an explanation for this: by walking the bike up a short but very steep incline at the beginning, we coasted gradually downwards for most of the remainder. We called that route our Moebius trip. It was one of my favorite low-tech tricks, like winding a clock and letting the stored energy in the mechanism do the rest.)

One stop on our route was Edna's. Edna, the Minimite midwife, was married to the deacon and lived on the opposite edge of the community from us in a remote and woodsy cul-de-sac. She was soft-spoken but self-confident, with a steady, calming gaze.

After a gentle pat-down of Mary's abdomen on the first visit, she determined that the baby was properly positioned, head down and posterior up. She recommended that Mary drink plenty of raspberry-leaf tea in preparation for delivery, to reduce chances of bleeding, and she supplied us with a large bag of home-picked raspberry leaves at nominal cost. She also handed us a kit of supplies that would be useful when the time came, with such things as plastic bed protectors in it. She wanted us to keep visiting until the ninth month to make sure the baby's position stayed the same. Finally, she asked that we arrange an appointment with the local doctor who backed her up; he would do an exam to make sure Mary was a good candidate for home birth. We arranged to have the exam, and much to our relief Mary passed. We were ready, it seemed, for birth in the comfort of our own household.

Edna didn't charge for any of her services but did accept voluntary donations. With nothing to give her that she needed, we resorted to fungible currency in this transaction. A hundred dollars was more than she'd ever gotten, she said. Fair enough; in return we'd receive more than we'd ever gotten too.

THIRTEEN

secrets and politics

I had attended church services, barn raisings, and hoeings, but there were other Minimite meetings that no one talked much about and that I had never visited. One time when I dropped by Sylvan's, he wasn't home and I waited for him for a few minutes. He finally pulled up in his buggy, and I asked him where he'd been.

"In council," he said.

"What's that?" I asked.

"I'm not really supposed to talk about it."

The answer only piqued my curiosity. I gathered from his secretiveness that the meeting was unusually important. I had to know more. I gave Sylvan a hurt puppy-dog look.

He sighed. Finally he began spilling some beans. He couldn't get too specific because personal privacy was involved, but he could reveal some generalities. This information, combined with tidbits provided by others, helped me piece together the puzzle.

The council was the Minimite form of self-government. It was their apparatus for reconciling differences and choosing technology. It was also the supreme test of *willingness,* or self-surrender: only by first yielding power to one another and to what they regarded as the spirit of their unity, a spirit from above, could the members wield it over machines.

The form of the council was not rigidly prescribed but developed, more or less *sui generis,* along topics and thoughts as they arose. Like the religious service, it was contemplative and querying, punctuated by periods of pregnant silence.

After an opening prayer, followed by a pause for reflection, members one by one would speak their minds. Older members went first, followed by younger, until everyone had had a say. Formal votes were not taken. Usually the general sentiment was clear enough, or if not, the ministers—chosen by popular nomination followed by a lottery—decreed the outcome.

Max Weber, somewhere in his *Protestant Ethic*, drew a sketch of an Anabaptist meeting and made it clearer why the conclusions reached in such gatherings could be so penetrating:

"The peculiarly rational character of [Ana]baptist morality rested psychologically above all on the idea of expectant waiting for the Spirit to descend, which even today is characteristic of the Quaker meeting . . . The purpose of this silent waiting is to overcome everything impulsive and irrational, the passions and subjective interests of the natural man. He must be stilled in order to create that deep repose of the soul in which alone the word of God can be heard . . .

"But in so far as Baptism affected the normal workaday world, the idea that God only speaks when the flesh is silent evidently meant an incentive to the deliberate weighing of courses of action and their careful justification in terms of the individual conscience."

Put another way, Minimite councils depended on a secret similar to the one Mary and I had discovered while relaxing with the windows open in the evening: in true leisure there is mastery. If the enemy of self-direction was passion and impulse, its ally was quiet repose, mindfulness, perceptivity. Yet the act of reflection transcended the rational; it followed a course that could not be entirely foreseen, yielding conclusions that could not be reached if too deliberately pursued. From this quiet, something surfaced—a reserve store of living experience, a world of subtle connections and insights, and possibly (as the Minimites held) inspiration from other more inscrutable sources.

As it happened, the subject of the meeting this day, Sylvan divulged, was "a little chat on the relation of matter and spirit." What he meant by this, more literally, was the extent to which human causes and plans need yield to Providence. The talk was not abstract speculation. It had a concrete focus: the telephone.

Personal telephones, as I already knew, were not allowed in Minimite practice. But pay telephones had never been off limits, and their use was creeping in. No one disputed the need: a midwife had recently called the doctor to save a mother's life during delivery. But members increasingly were using pay phones—and their neighbors' telephones—for less urgent reasons, like setting up business transactions with produce wholesalers. They were starting to drive their buggies great distances to reach convenience store phone booths.

The exchange went back and forth. Frequent use of the phone, some felt, encouraged idle chatter, sped the pace of life, and reduced real-life contact with members of the community. But banning phone use altogether, others countered, was out of the question. No one wanted laboring mothers to die for lack of a doctor. In the end a compromise was reached. The council agreed to continue allowing use of phones, but only when urgently necessary. At the same time, they began to make plans to coordinate produce sales using a single Minimite go-between.

As news of this discussion sank in, I realized there was something astonishing about it. Here were members of an obscure sect in a prayerful meeting—rationally evaluating the implications of a technology that the rest of us accept on faith. You could even say they were in the process of conducting an *experiment* on the telephone. Would that the inventors of our gadgetry were so scientific.

The mood during this exchange remained collegial and focused, but when the topic veered to another matter, a rift opened up. The use of telephones, it seemed, was not the only source of hurrying and scurrying. Another was land speculation. Members who had migrated from Lancaster County were buying up ground as fast as they could out of fear of a future shortage. They were used to the high prices and scarcity of farmland on the east coast and wanted to make sure their sons would one day have their own farms. But this defensive buying strategy only drove prices up and vastly increased the work needed to pay for and maintain farmland. Minister James had preached on a Gospel parable about a man who refused to pause for refreshment because he was too busy buying land; among the Minimites the tale was not fiction.

When it came to choosing sides in the debates, Sylvan made no bones about the way he leaned—toward Providence. "Consider the lilies of the field," he said, "how they grow, they toil not, neither do they spin!" Others, however, lent their support to the cause of hard work and ever-bigger family farms. Sylvan gazed heavenwards. They were returning to the rat race they had come here to escape.

He could not conceal the revulsion he felt toward this defensive land-acquisition strategy. He suspected that behind the practice lurked a simple lack of faith. But he did not hold out much hope of dissuading the land buyers. They were too set in their ways.

The standard-bearer of this faction, besides, was one of the most respected and articulate men of the group. He was a likely candidate for a new ministerial position that soon would be created when the growing settlement divided into two districts. He could become, effectively, the head of the whole community. If he were elected, its identity might at last become defined—at the expense of a certain reasonable breadth. The name of this person was Edward Pendleton.

The news filled me with foreboding. What troubled me most went beyond the mere issue of land acquisition; it was the way Edward had made this *particular* piece of land, this *particular* community and code, the only answer. Mary and I didn't want to make the same mistake. Until Edward's role in local policy became clearer, perhaps it would be better not to narrow down our options, not to put all our eggs into one basket. It was the principles of the Minimites that we hoped to enshrine, not the Minimite community itself. A principle is not the prisoner of the particular. It is transportable.

But could we stomach another trip? Our first scouting expedition had produced the meagerest of results, but we had been rather picky about what we had wanted. Perhaps we were too particular. Maybe we needed to be more fluid, more open to the unexpected. Maybe we needed to think of *Gelassenheit* in a new way.

My mother was aware that we were scouting out sites, and in one of her letters she suggested a spot near Topeka. My first impulse was to snicker. Apart from the fact that I had no nostalgia for my home on the range, Mary and I weren't considering moving any farther from her part of the world than we already were.

But I had to admit we had come up with little in the East; and by an odd chance, mutual dear friends from Boston had recently relocated near the place Mom mentioned. It seemed we could combine an exploration and a reunion and appease my mother all in one fell swoop.

We threw a few pieces of luggage in the car and took off. I must have driven by the highway marker many times, but I had never had a reason to turn in. When we left the interstate, we descended upon a lush and rolling river valley. From a distance we spied the town. It sat on the opposite bank of a wide, churning river, nestled between hills and bluffs. The beautiful red brick buildings were ornate and spired, and many appeared to predate the Civil War.

As we got closer, I saw homes and businesses and public buildings all intermingled in a six-block-square area near the river. We already knew, after doing a little research, that there was an Amtrak stop and a bicycle trail on the bed of an old railroad line that had linked the town to other population centers in the state.

Peering more intently, we noticed goats and horses grazing in some tidy backyards. In the shops the people were bustling and garrulous. Traces of a German accent floated through the air from the many descendants of the original colonists, who, a brochure told us, were transplanted from Philadelphia in the 1830s. They were Germans from Pennsylvania, it seemed. But they were not Amish.

The difference was twofold. First, the major crop they raised was not corn or hay, but grapes. Second, the religious constitution was mixed and supplemented by a strong legacy from founding German "Freethinkers." Making wine was one of the chief occupations of the citizens, and enormous caches of it were stored in a catacomb under one of the hills in town. The long-standing heritage of this place, its sense of social cohesion, appeared fairly intact. The steep hills and the wine industry helped hedge it in and focus it.

The town had a Catholic church, a massive red brick monument with antique stained glass windows, set high on a hill in the center of everything.

A real estate agent showed us a charming house in our price range (about the cost of a medium-priced recreational vehicle) at the top of the hill near the church. It had a panoramic view, and the south-facing backyard offered a workable site for a year-round garden, using cold frames and plastic tunnels. The agent, who'd lived in the area his whole life, told me he economized by cutting firewood to heat his house. And he heated the water for the town car wash solely by wood. Wooded tracts of land were available outside the town limits, and he could sell me one, he said.

And yet this place was not so anachronistic as it seemed. It was modern in its utilization of everyday technologies. It lay just outside commuting distance from a major urban area. It had, in fact, nearly all the ingredients we were seeking. My mom's tip had not been half-bad. It was possible to picture smuggling Minimite principles into Modernity via such a place.

But there was a problem. The town had no college. What would I do for a living if not teach? Could I farm? But how could I take part in the life of the village if I did?

Did I possess other skills that I could use here? A thought came to me. Impulsively I approached an attendant at the visitor's center with it, a woman with brunette hair and a large open face.

Yes, she replied, I could *try* a horse-drawn cab service. Someone else had already done this. "But one time his horses bolted. They got scared or something when a big semi went by. I never found out if there was a lawsuit, but he stopped doing it. It really scared him."

The attendant put the kibosh on the horse-and-carriage idea. But Mary and I left this remarkable historic island deep in reflection.

FOURTEEN

heatstroke

When we got back from our trip, the midsummer lull was over. The season had advanced a notch. It was time for threshing, and the Millers asked me if I wanted to take part. I remembered a comment my grandfather had made that in his day threshing had been a test of manhood. Having undergone a three-and-a-half-hour church service in a stuffy room, I shrugged. If that wasn't a rite of passage, what was? I did know that field work was probably even more typical of the kind of shared labor the Minimites engaged in than barn raising. And in many cases it involved the use of technology they themselves had cleverly tailored in order to preserve the cooperative character of the work. I didn't hesitate to say yes.

In our absence a heat wave had moved in, and as I approached the threshing site on foot, the very air seemed to weigh down on me. Once-distinct hills and trees melded together into a gray-green smudge. Dust hung everywhere.

By the time I rounded the last bend in the road, I was beginning to feel lightheaded. The sight in front of me seemed almost surreal. The farmstead of Gideon Stoltzfus, like others I had passed on the way, consisted of an unadorned clapboard house and a tall metal barn with attached silo. West of the barn lay a pond shaded by a few trees, where some geese and pigs lounged in the mud. Behind the buildings were more trees, and off to the east there was a large bare field, recently cut down to stubble.

Everywhere bearded men milled about, intent on various aspects of this multifaceted task. When the man on the Greyhound bus had first alluded to his life without motors, I had assumed he and his neighbors collected the wheat with scythes and threshed it in the air like characters from Biblical times. How else could they obey the rules? Before me in the doorway of Gideon's barn sat the answer: a huge old contraption, an amalgam of galvanized metal and weathered wooden boards, quaking like an elephant with palsy as it swallowed the sheaves being fed into its snout from a trunklike conveyor belt. This was the threshing machine.

Stretching from a hub on its side was a black rubber belt that reached about fifty feet to another piece of machinery in the middle of the barnyard, around which circled four pairs of horses. At some point I found out that Gideon, who co-owned the water mill, had put together this device with the help of his other brother, Elbert, using the discarded transmission of an old dump truck. It was a horsepower rotor.

Despite the use of these non-motorized devices, there was still plenty of work for humans to do: two men pitched sheaves onto the conveyor belt from a high-mounted wagon; two others tended to horses held in reserve in a shaded enclosure near the barn; many unseen were gathering sheaves in the field; and a small boy stood rotating in the center of the turnstile, prodding the horses with a long stick.

To me, the scene was as strange and exotic as a three-ring circus. And to think I myself would soon be playing a role in it.

An older, portly man with a long white beard nodded his head in my direction. I recognized him from the church meeting: Bishop Henry. "Eric?" he called.

We shook hands and made small talk for a few minutes. I was taken by his solicitude, which appeared genuine; but our conversation was cut short by the departure of the harvest crew. He put a pitchfork in my hand and pointed me in the direction the other men were heading.

Following them to a flatbed wagon, I scrambled aboard as they did. Another boy, standing at the front, flicked the reins and two Percherons sprang into motion. In anticipation of their forward lunge

I was already sitting cross-legged on the wagon bed while the men stood and swayed as it undulated, like sailors on the deck of a ship.

After a short journey behind the barnyard area and through a hedgerow, we came to a field littered with shocks of wheat. The tumbledown oat piles Bill and I had put together could not compare with these. Each was a small Parthenon, precise in the number and spacing of its fluted sheaf-pillars which supported a perfectly symmetrical sheaf roof.

The task before us almost seemed a desecration. After spilling to the ground, the men somberly and silently began thrusting their forks into the first shock. In moments it no longer existed. They moved onto the next one, and the wagon followed.

As I looked closer I observed that they were catching the sheaves (the bundles of wheat of which the shocks were constructed) by their twine bindings. Thrusting into as many as they could hook in this way, they would lift them high over their heads, balancing them aloft on the short trip to the wagon. There they dropped them. Another worker who stood on the flatbed scampered to arrange the sheaves, packing them tightly in layers.

The time had come for me to give it a shot.

I began by impaling a single sheaf. It was surprisingly light and I wielded it easily, thrusting it upwards, balancing it overhead as I stepped toward the wagon. There, imitating the others, I gave it a gentle flick. It landed right where it was supposed to, and the packer tucked it in place. There was something delightfully gastronomic about the task as if, with my long fork, I were placing oversized biscuits of shredded wheat onto a platter. As my confidence increased, I upped the helping to two biscuits.

As the pile grew, I will confess, it was no longer so easy to flick the sheaves to their spot at the feet of the packer. It took a healthy grunt to place the load onto a rising mound that was now at shoulder height.

. . . Now it was over my head.

. . . My right arm began to burn.

A sheaf I tossed came back on me and hit me rudely in the face.

A couple of the men smiled.

The size of the heap on the wagon seemed tremendous, yet the crew kept adding to it, thrusting up sheaves as if they would never tire. Occasionally some mopped their brows and remarked on the day's unusual warmth, but most remained stoically silent. I tried to muffle my grunting sounds—I fear I might otherwise have roared—as I strove to hurl my sheaves ever upwards.

At last someone muttered, "Call it a load."

The straw Matterhorn rolled slowly away, and an empty wagon took its place.

Gradually the rows of golden Parthenons fused into an immense bonfire shimmering in the gelatinous air. My temples began to throb, and everything started to get blurry. Only the occasional, slight movement of air brought relief; my clothing was like a plastic bag that sealed in perspiration. The heat was like nothing I'd ever experienced. It seemed to bake me from within.

One of the men turned to me: "How's the 'good life' treating you?" There was a smirk on his face. The phrase "good life" rang a bell . . . I realized he was quoting from my letter. Had the bishop shown him my letter? Had they *all* read my letter? Blushing (as if more red would show on my already crimson face), I tried to gauge my response. Fine particles of straw dust permeating the air chose this moment to gather in force behind my nostrils, however, and instead of saying something, I sneezed. Sneezes became sneezing attacks became paroxysms.

Someone standing above the scene during the next round might have beheld what looked like a thresherman playing blindman's bluff. I have dreamlike recollections of a wagon loader dodging my pitchfork. I can still hear a voice saying, "When you start to get chills, you know it's time to stop."

Perhaps it would have been better if they had carried me off the field, but I stuck it out to the end of the second round and then rode with the crew to the barn. There, next to the conveyor belt, dismounting, I met the bishop. He looked me up and down, furrowing his brows as if beholding a baby bird that had fallen from its nest.

"Are you feeling all right?" he asked.

"I'm fine," I wheezed.

"You know, you don't have to continue."

"I'm *fine*."

"Well," the bishop went on, "if you're *sure* now. There is another job we could use your help on. But only if you want to." He led me into the barn. We had to squeeze sideways to get past the metal elephant, which filled up most of the doorway, but once in a back area came to an enclosed space in which the machine's discharge flap rose and fell, like a flopping tail, to make way for the excretion of digested grain. I was at the very bowels of the beast. A small boy was fitting grain sacks around the tube and lugging them away as they filled.

For a couple more hours I took turns with the boy filling sacks—a task that in my fogged-over imagination bore a strong resemblance to the removal of elephant turds. My sense of smell was inoperative, precluding minor distinctions.

When at last I felt I had made an adequate showing, I excused myself. My head felt about to split open, and my skin was as dry as parchment. I wandered home dazedly, like an astronaut on the surface of the moon.

I was bedridden for three days, tossing and turning in a feverish delirium.

—

The only thing worse than unconsciousness was coming to. I sat up abruptly with a chilling realization: threshing wheat without modern equipment had shocked me out of my senses. It appeared to be a form of naked physical toil that only a glutton for punishment would willingly engage in. Was there a need for more machinery here?

With the throbbing headache I had now, it was hard to recover any enthusiasm for life without motors. Maybe it would be easier, until I had time to think things over, for Mary and me to rent a sports car, escape to San Francisco, and eat sushi by the Bay. For that matter, I could take up a career in Silicon Valley, commute to work, and forget about ever returning to M.I.T.

My mind began circling back over the events, again and again, finally coming to something that didn't fit.

The heatstroke had been selective. The other workers had come out okay. It had struck only me.

Why not them? Why me?

I was comparatively out of shape, yes, but this had been their first day in the new threshing season, too. Was I *so* physically unfit?

Then I remembered the week of traveling. When we weren't riding in a car, we had been sitting inside air-conditioned buildings. Meanwhile, back home the temperature had steadily risen, and presumably the Minimites had adjusted along with it. I alone hadn't. If anything, I had made a reverse adjustment.

I checked my hunch in a medical guide. The body, I read, normally takes two weeks to adapt to major temperature changes. This is a physiological process, altering blood-flow patterns and increasing the capacity to sweat. Any too-sudden introduction of hot or cold can overwhelm adaptive mechanisms.

The book went on to explain why I had become so weak in the heat. Because I was unable to sweat adequately, my circulatory system tried to compensate by rerouting blood from my body's core to the skin and extremities, thus carrying oxygen away from the muscles. My muscles simply ran low on fuel.

But the failure of my cooling system was caused by interference from an electric one: the air-conditioning had weakened me.

But if threshing *per se* did not refute the possibility of Minimatic convenience, then something else appeared to do so: heat.

After getting back on my feet, I still quailed at the heat. If quiet magnified the possibilities of free time, heat shriveled them. Heat rerouted blood flow to the skin, draining the brain. Contemplative serenity wilted in the heat. Heat drove Mary and me to distraction—and to mutual uninterest. It even discouraged sleep. There was no rest even during our rest.

The sleepless nights and still humidity were especially wearying. It was Mary who finally had the insight (slow in coming, not surprisingly). Our bedroom had three large windows, but the bed itself was pushed against an interior wall, removed from the ventilation. Mary simply took the pillows and placed them at the foot of the bed. Now we caught the cross-breezes. They were sometimes slight, but even so it made a terrific improvement. The change could hardly be simpler, and partly for this reason easy to miss. The whispers of wind would come in slow, rhythmic breaths like the hand-strokes of a friendly ghost . . .

The air . . . was . . . conditioning . . . 'us.

As we moved toward midday, the breezes grew stronger, but so did the sun. The wide straw hats we had taken to using provided some relief, but not enough. As I hoed the garden on one still, sweltering day, I began to sweat profusely. I could feel the sweat pouring out and felt helpless to stop the flow. I tried to put the sensation out of my mind. My shirt clung to my back and slithered across my skin. Perspiration trickled down my forehead and into the corners of my eyes, blinding me.

I escaped indoors for relief, and then something strange happened. Under the skin, up and down my arms and legs, I began to feel millions of tiny electric jabs. I'd never felt anything like it. It was like some sort of subcutaneous massage. Staying still heightened the feeling. Perhaps my movements in the garden had prevented me from noticing until now. I paused a while longer, enjoying the strange sensations. My body was finally coming into adjustment. It was learning to sweat.

Mary remained indoors or in the shade more often than I and did not need to undergo quite so dramatic a transformation, but she too gradually acclimatized.

After about five days, I felt completely recovered, and with some trepidation began to wonder whether I might be ready to make another go of threshing. Only one way to find out. I decided to bring a small pocket notebook this time to record my progress and any other details of the occasion as they unfolded. I realized the attempt might draw attention to itself, but I hoped to be discreet.

The event this time was to take place at Alvin's, brother to Gideon and co-owner of the water mill, and when I arrived I found a half-full wagon and a crew in the midst of work. Surrounding me on all sides, again, were those wheaten Parthenons. Brandishing a pitchfork, I set into demolition, catching an earful of the conversation already under way.

"What, you use insecticide on your potato bugs?" Elbert was saying. "Why, I jus' let the neighbor's guinea birds take care of mine. Whomp, whomp, whomp." He lifted up his nose and chomped like a guinea sucking down a potato bug, his long beard flapping.

"Good year for insects, with all the rain," came a voice from the other side of the wagon, an invisible thresherman concealed behind the ever-rising mound of sheaves we were creating.

"Yeah," added Gideon with a groan, as he thrust a sheaf high on the pile. "Last year it didn't rain and nothing grew. This year it rained and nothing grew. Guess what I got from my blackberries the last two weeks? A big fat zero."

But then his eyes tilted mischievously. "The weather," he continued, "it always seems good for something. [Grunt.] This year, it's strawberries. The children, they're getting a little tired of them now. Wish the truck folks would take 'em off my hands. They don't want 'em. They want this other stuff. [Grunt.] The only time I ate it, I had to close my eyes. But it has"—his face lit up—" 'shelf-life.' " He could see us grinning and knew he now had center stage. "They don't want our King Crimsons. They want this other stuff. Ours is perishable. Theirs has"—again the bright face—" 'shelf-life.' "

Gideon had a large and fascinating nose. It was not grotesque, exactly, but somehow beautiful, wending and bending and calling to mind a cornucopia. He turned his head one way and tilted his eyes the other, smiling demurely as he chattered. Thoughts and glances, winks and gestures, were woven together in an intricately coherent counterpoint.

"Now you take watermelons," he warbled on. "I like 'em with the skin this thick." He pinched his fingers together. "But they won't take 'em. You try to ship it and"—he squawked as the food shipper would—" 'it cracks!' They want something with skin an inch thick that they can throw around. We pick fruit that day and have it for supper. They pick one day, pack it the next. They pick it green"—Gideon's voice cracked as the other men broke out in laughter—"ship it the next, and sell it who knows when. Well, we really live like kings."

Silent until now, I at last offered my own observation: "It never occurs to people in the city that, if they lived in the country, how well they'd eat."

"But they wouldn't want to do what they'd need to in exchange."

Added the invisible voice: "They don't want to have to work." An audible grunt followed, and a sheaf rolled into place on top of the stack from behind the mound, nearly completing it.

Work . . . That's right, we were working now. I had almost forgotten. Nice that I seemed to be keeping up the pace, barely feeling it. The temperature today, admittedly, was a degree or two lower than it had been that other time.

Maybe I'd go another round.

As I began pitching sheaves once more, that voice from behind the wagon inquired: "Gettin' too tired yet?"

I realized he was talking to me. "No, not at all," I answered. "Not making the mistake of taking three jabs each time instead of one." I demonstrated by scooping up a couple of sheaves in a single sweeping motion, which I'd copied from the other men. I did feel a slight, pleasant ache in my thighs, back, and shoulders. Since my last stint, it was almost as if my body had developed a craving for work. Odd, that pain could feel almost . . . good.

"Yeah, we remember," guffawed Gideon. "Your face was so red, we didn't know what to think about it."

Before I knew it we were back at the barn again, fetching an empty wagon. There Gideon's father, Levi, the bishop, came upon me. "With your light complexion," he inquired solicitously, "is the sun beating too hard on you?"

"No, I'm all covered up by this hat. The only thing sunburned is maybe the back of my neck."

When we got back to the fields (round three), Gideon probed a little further. "Do you think you'll sleep well so tired, or maybe all the aches and pains will keep you awake?"

"If I feel tired, it will be tomorrow morning when I try to get out of bed."

Gideon laughed. "No, I don't have a problem with that. But I do feel tired at the end of the day. But one time, well, we were building a roof shaped like this." I turned around to see his hands folded into an 'A.' "And I had to hang from a rope like this"—he bent at the waist until his body formed a right angle—"all day long in one position. Boy, was I sore the next morning. Then I finally got to know what it must be like to be an old man."

Gideon provided good material to demonstrate the self-automation of manual labor because his very commentary on it proved he wasn't thinking about it. The more absorbed he became in his stories, the more apt he was to forget where he was or even who he was. It was as if he inhabited his subject psychically. He was a real raconteur. Whenever he looked your way, you felt a sense of unmerited privilege as if the attention that might have been lavished on a large audience was focused on you alone. You were the reason for it all; you were his public.

He began recalling exploits of the logging industry near the place where he had grown up (he and his brothers had also relocated here to escape development), and somewhere in his account he began to identify so closely with the trees, it was as if had taken their point of view. Next he stepped into the mind of the lumberjack, the busy worker clear-cutting the forest from the mountains. He kept repeating the term "clear cut" as if he were a rapt believer in the practice. His allegiances kept shifting, but as a storyteller, it was his gift to see things from all sides. Now with pain he began to recall backwoods people throwing stones at him or trying to exact tribute payments the times he (now Gideon again) tried picking berries from the brambles that had grown over the bared mountainsides. Next he took the point of view of the forest ranger monitoring the fires people were setting to clear the way for fresh blackberry growths. What if *you* had set one of the fires? "If they don't see you go in, they won't know who did it, but if they see you go *in*"—he made a dramatic pause—"they have an *idea*." Gideon the forest ranger now pondered the thought of closing the cuffs over your wrists.

It was happening as it had before. Physical motions were becoming automatic, freeing the mind for other interests. Muscle fibers meshed, synapses branched, the heart pumped, lungs swelled, and sweat glands—now at full throttle—discharged. Meanwhile, I listened to Gideon. When the dinner bell clanged, I was actually disappointed to have to climb from the wagon.

If the body shares characteristics with an automatic machine, then one of these is its need for fuel. But just as the body is much more than a machine, so is its food much more than combustible fodder. Instead of gasoline, we set our sights on a banquet-size table laden with home-made mashed potatoes streaming with fresh-churned butter; trays of tender meats; loaves of fresh bread that pulled apart like cotton; fresh greens and vegetables from the garden; bowls of cup cheese; sweet gherkins and berry preserves; cake and fruit crunches topped with fresh cream.

Lunch was a culinary gala, an intense communal event mixing art and nature, as threshing had been. Toward the end of the meal, after most of the men had had their fill (but I still had half a plate), one began to recount another story. The day before, a customer visiting his roadside stand had insisted on " 'eggs from a hen.' The last time he bought some, they"—munch—"came from a carton!"

The room rocked with laughter.

"I heard a news item," said someone to my left, "that they want to make it illegal"—munch—"to eat eggs in some places."

Everyone chorused, "Eggs?"

"Too much cholesterol. They said"—munch—"the average American doesn't get enough exercise to burn it off. They're innerducing legislation"—munch—"somewhere about this."

More raucous laughter.

"They ought to make it illegal not to exercise."

"I bet a lot more people die"—munch—"from that tubacka every year. I'd like to see the status ticks on that."

"What if everyone was just given enough ground"—munch—"to grow their own food on? Then"—munch—"these lazy people would have to work!"

"It won't"—munch—"happen."

Conversation circled around the table, and before I knew it I had finished my dessert. I looked at the cleaned plate and frowned. I hadn't even tasted the last two courses.

That automation of physical functions can cut both ways.

On the wagon after dinner there was a little flatulence, a little belching. We couldn't contain *all* the emissions here.

—

When the temperature exceeds a hundred degrees, even the most elegant bodily mechanisms begin to approach their limits. Mary and I had heard about a water hole somewhere near Nate's "Pa Kettle" homestead, and promptly set out for it. On reaching his place, we found him hitching up his wagon and loading up his children. He had the same idea.

We joined about five of his children on the flatbed and bumped across a pasture full of cattle, then down a terrifying embankment towards the creek. (Afterwards I realized thankfully that Nate must have installed his brakes.) After tying the horses to a tree, he led us along a dry creek bottom for several minutes. Then we came to the swimming hole.

I say "hole." Nestled in and alongside an exposed rock face, with ledges in place of diving boards, lay a pool so magnificent one could only pause and gape. Ancient craning trees shaded the mirrorlike surface, which reflected back depths of green foliage. The trees, the water, and the rocks formed a kind of *faux* cavern, a den of terrestrial refreshment. Nate's children darted ahead of us and jumped in; we set down our towels and books and followed them—*kerplunk*—into the purest, darkest, most delicious coolness. The restoration was complete and all-enveloping. After numerous dunks, we got up and romped around with the kids. We made flying leaps from the rock ledge. We shouted with glee. We splashed like seals. We finally sank on our towels along the gravel bank, positively shivering in the breeze. When we finally dried off, we jumped in again, swimming until the light began to ebb in the late afternoon.

So did we learn about another natural alternative to the air-conditioner.

There was still another.

Late one sultry afternoon a big storm moved in, tossing trees and carving out gullies with rivers of rainwater—and rinsing away the humidity and heat. The next day, it was about seventy-two and clear. I was shivering again.

It occurred to me that this new coolness would not be nearly so bracing had it not been for the unbearable weather before.

In our era of high technology, affluent westerners spend billions every year to "get away" to exotic locales. They do so surely to escape the stress and frustration of modern life, but also to relieve its monotony. They spend forty-eight weeks of the year in the same job in a climate-controlled environment; when they go home in the evening, they travel on the same stretch of freeway to a subdivision where all the houses look the same; they watch television programs that reduce the complex issues of life to half-hour segments on a flat screen. They crave diversion, depth, escape. So they fly to Bermuda. Or for a few precious days, they stroll through Disney World's mockup of the architecturally diverse midwestern downtown their grandparents once ambled through whenever they wanted to, and spend all the money they saved during the previous forty-eight weeks in the same job.

There may be another way. What if they just noticed the weather changing? Those who lack western affluence already rely on the weather for daily variety. New weather alters the look and feel of the landscape without altering your location. You don't have to travel elsewhere to experience the exotic; the exotic travels to you.

A day or so after the storm, the features of the landscape stood out clear and green. There was a stiff, dry breeze. Big Sur, seemingly, had traveled over to the Midwest. When I joined the threshing crew, it was indeed as though we were somewhere we'd never been. The crew had a ruddy, windswept look, and our straw hats kept blowing off despite our best efforts.

On a chilly morning after the rains, I woke up to find mists rolling forth from the heavy dews like roiling smoke. Half the sun poked over the horizon, intensely orange and perfectly circular like an egg yolk. Several layers of nimbus clouds hung above it like the bough of an enormous tropical tree, or like an outstretched hand with the lower fingers curled under.

Grace told me once: "If the sun rises and hides itself soon, then rain before sundown, and may before noon." The sun didn't quite hide itself that day, but as if undecided, the weather remained ambivalent. A strong, cool northern breeze brought scads of differently toned clouds. Some were white tufted cotton balls, some hideous purplish things, and when the sun poked through, they lit up brilliantly. Later the weather became more organized, with the white clouds staying to the right and the purplish ones to the left. In between there was a bright gap, through which a cool, fine mist descended.

Sometimes great weather differences alternated in very short intervals of time or hovered only a few feet apart. On a very breezy day, Bill and I were picking cantaloupes together when out of nowhere the wind changed direction. It had been coming from the south, but now a northerly cloud bank rolled in. A cool breeze splashed against us, swirling and licking through our hair. Then the wind shifted to the west. Bill, who was about fifty yards away from me, exclaimed, "Gee, it's warm over here!" I was still relishing the cool breeze. For a minute or two, the winds vacillated; sometimes Bill was warm, sometimes I was. It made that much difference fifty yards apart.

Another time when I was working alone in the garden, the morning began gray, cool, and breezy. I was dry and cool in the breeze. Then the sun came out from behind the clouds and the breeze quit. Suddenly I was sizzling. The back of my shirt became drenched with sweat. Then the sun disappeared again, and when the breeze started, I got goose bumps all over. So it went in ten-minute intervals all the morning. I was in heaven.

My favorite weather juxtaposition, though, happened one afternoon when I was riding in a buggy with Edward along a ridge. It had been hot and still, but sudden gusts of wind began to assail us. Though the sun above was extremely bright, dark gray masses were stealing along the northern horizon, tickling the ground. A ferocious storm was nearing. It was as if another whole kingdom existed just over the hill, a kingdom of darkness that carried its own principles of organization with it, a different kind of matter and space. Then its weird coolness began to swirl in little eddies around my ears. The storm seemed to be moving in the same direction we were, and for a while my skin prickled as I savored the journey between two king-doms, in the delicious position of one who can touch one world on his left, another on his right, and be a part of either.

But of all the weather displays, none could compare with the sun-sets. There was an evening after the rain had fallen. In the hush of dusk, dozens of birds began to chirp sweetly like a chorus of merry sprites. The cool, moist breath of imminent nightfall moved across my face through an open window. Muted white clouds were softly embedded in the blue ground of sky. And there, visible behind the trees, casting its yellow light behind the barn and onto a sapphire expanse of young corn, was the fading sun. The sun itself, though, was not so remarkable this time as its effects.

Another time it was the other way around: the scene framed the sun. The clouds were large white fluffy castles moving slowly through the dusk, pierced through with many-shafted light. The sun was like a prisoner, looking out from behind crenulated openings, cracking flares, trying to make itself seen, its large eye roving from window to window.

Weather can be two-edged in another way: It may stir up feelings of melancholy or memories from the past, take you to a place you do not necessarily want to go. Or it may return you to where you started—which in some cases is another way of saying the same thing.

weighing the work

Just as surely as cool air wafted in, warm, muggy air took its place.

It was James's turn again with the threshing machine. Other crew members climbed aboard the wagon with me: Elbert, swarthy and broad-shouldered; Wilbur, bespectacled, lanky, and loose-limbed. Wilbur was someone I had yet to get to know, but whether it was the effect of his glasses or his lankiness, he was the first Minimite I'd seen who seemed unsteady when the cart began to move.

As we hoisted sheaves, Elbert asked me point blank: "So how hard do you think the work is?"

It was an opportune question, and I thought about it in light of recent changes. It surprised even me when these words came out of my mouth: "It all depends on your mood. The most beautiful job can seem horrible, and the most horrible job, grand—depending on how you're feeling that day." Here I suppose I meant to include all the factors that contribute to a mood, such as weather and one's ability to respond to it.

Wilbur's eyes grew big behind his spectacles. "I would agree with you on that," he said. "I would strongly agree with you on that." He spoke in a soft, breathy baritone and gazed at me with large, sad eyes. The others continued to prod me about how I was taking the heat. I rewarded them with harrowing accounts of might and main on the Yale rowing team: how I'd undergone twenty minute periods of weight lifting without pause or rest, relay races up ten flights of stairs, rowing in arctic temperatures . . . heroic feats of modern fitness past.

Wilbur eventually cleared his throat, and with labored politeness began, "But for a change of subject, I would like to go back to what we were talking about before, about moods. Do you think that there is anything you can do about moods?" This question pulled the conversation up short. The men, now silent and wary, continued to hoist sheaves. I released mine (there it was on the end of my fork) onto the wagon. Wilbur, I sensed, was talking not so much about moods in the abstract as about himself.

"That depends," I answered, "on the kind. With some you can, with some you can't. A lot of the time, it's just a matter of accepting the mood itself instead of thinking there's something wrong with you. If you get up and feel sad, rather than pretend you're happy or act as if it's terrible not to be happy, you should say, 'I'm sad today,' not act as if you were always supposed to be happy. Otherwise you really tie yourself in knots."

I went on with my amateur psychobabble, describing studies I'd read on depression among the Lancaster Amish. There were two kinds of depression: "unipolar" and "bipolar." The first was everyday, "common cold" depression; the second, a more serious organically based malady, sometimes called manic depression. Despite all their modernization, the Lancaster Amish still largely sidestepped the American plague of "unipolar" depression; they were five or ten times less likely to come down with it than the general population. "Bipolar" remained equal among Amish and outsiders.

"How can you tell them apart?"

"I really don't know exactly, but a psychiatrist could do it most likely."

Wilbur looked thoughtful, but from here the conversation drifted. The shocks seemed to grow heavier and heavier. We were into oats, and oats, someone said, are half again as heavy as wheat. Guesses at the temperature ranged from eighty-five to ninety-five degrees—it was hard to tell because the humidity was so high. Every sheaf felt like sheer, dead weight—like a bowling ball on the end of a stick. Inertia was all that kept me going. Objects again began to swim hazily before my eyes. Conversation took conscious effort, like the work itself.

It was, frankly, a little depressing.

As the moving mountain of sheaves retreated to the barn, my eyes shifted to two tiny figures in the distance bearing an object that dangled between them. The men tensed in anticipation. The figures enlarged and became teenage girls in white head coverings; the object became an ice chest. Up flew the lid.

Ice cream!

In three large translucent plastic pails, raspberry, praline, and vanilla. Cool, foamy, rich, and sweet, it settled in the stomach, radiating a heavenly chill throughout the torso and upper abdomen.

Last winter's pond ice, preserved in sawdust, had come to the rescue.

Rejuvenated by the ice cream, conversation rebounded. Did you hear the latest? Someone had told so-and-so *personally*. You, yes you, can get a pet pig that won't grow larger than forty pounds. It comes house-trained and costs five hundred dollars.

"Who would want a pet pig?" somebody asked. "Pigs are the *dumbest* animals."

"Pigs," I corrected him, "are smarter than dogs."

Wilbur shook his head and snorted. "Someone told me that too, and I said, 'The only thing dumber than pigs is the person who says that dogs are dumber than pigs.'"

Squinting to avoid the suspicion in their squinty eyes, I stood my ground: a pig, I pointed out, recently rescued its drowning owner from a pond. Pigs outscore dogs on I.Q. tests.

Wilbur upbraided me: "How can a pig take an I.Q. test?"

A pig's I.Q., I explained, is determined by its ability to negotiate a maze.

We were now in a long and narrow field that dipped precipitously into a hollow, then rose up again.

Stories of valiant pet dogs began to circulate among the men, who swaggered around the wagon like sailors drunk with boasts. Wilbur, though, speculated quietly. "You might have something," he said, reverting to the humble tone of his religious upbringing. "Imagine if you raised all your dogs together in a pen the way you do pigs. You have a point there."

Alongside the oats ran Gideon's cornfield. He gestured toward it. "Achh!" he cried. Summer rains had prevented him from weeding. Tall misshapen redroots, lamb's-quarters, and thistles were sprouting among the rows like vegetative alien invaders. It was an embarrassment. But the others had similar weed thickets; each maintained his was worse. They began to turn over attack strategies: weeding by the buddy system, night hoeings, extra cultivation. (The community had long viewed herbicides as unhealthy for the soil.)

"They make good fertilizer," said Wilbur. Everyone stopped talking and looked up. He chose his next words carefully: "One time I didn't get a chance to stop them in an open field and disked them under. They turned it green just like it is after commercial fertilizer."

The comment brought a hush. A proposal to increase yields by reducing work? The thought seemed somehow illicit. "But if the seed gets mixed up in the feed," countered Gideon, "it goes through the livestock and comes out in manure, and then you're just taking it up the length of a football field"—he gestured toward the bare wheat field before us—"and sowing it back in. Those redroots, you miss one, it gets about twelve foot tall and, come fall, goes to seed. Seeds thousands or millions of new plants. Even if you plant in the fall and miss them, they'll grow four inches high and go to seed. They'll outsmart you."

The rebuttal seemed cogent, and the others waited to see what Wilbur would say.

"Sounds like they have a high I.Q."

There were several ways to take this comment. But from the wry purse of his lips, I gathered he was siding with me after the debate about pigs. I say this because there was also the slightest hint, in his wryness, that after all, next to a crafty redroot, a man whose mind is held bound by customs of the past is no intellectual match. Despite the sultry weather, despite the weeds, despite what may have been a bad internal disposition, intelligence had poked through and made it all worthwhile. What if we hadn't gotten that ice cream?

Even Wilbur seemed a tad cheerier.

That, by the way, was how I became acquainted with the community's other choice for the ministry, the contender certain thoughtful members would pit against Edward.

• • •

Wilbur's witticism, I will say, and his curiosity about moods got me back on track. Reclaiming happiness from the jaws of technology took more than a physical abandonment. It also entailed a certain pliability of mind, what Minister James had called *willingness*—a disposition to bend with needs and opportunities one could not have predicted. From this perspective human mastery is truly a mystery, a puzzle whose first clues may be honestly unpleasant or weigh us down merely because we do not understand or expect them.

In his own mysterious way, Wilbur revealed a certain capacity to understand this. His abashment was a bit deceptive. I think he was more in possession of himself than he let on.

—

But there was surely an objective side to the human experience with technology (or its lack), and technologists would gladly tell us so. One of the most exacting of them was the early twentieth-century efficiency expert Frederick Taylor, father of "scientific management." Wielding a stopwatch, he would measure the time it took a worker to perform a given task, such as shoveling dirt. Then he would analyze the task, breaking it down into segments, eliminating any unnecessary motions and replacing them with more efficient ones. The task was now standardized. Using Taylor's findings, a manager could instruct an employee how to shovel dirt in one perfect, unvarying pattern, as if he were a robot, and reprimand him if he deviated to the slightest degree. Taylorism thence became one of the most slavish forms of technological servility, parodied by Charlie Chaplin in the movie *Modern Times*.

After threshing with the Minimites a few times, I thought of a way to turn Taylorism on its head and call into question mechanically inspired ideals of human efficiency. It was never my aim to debunk efficiency itself, the effective arrangement of means to ends. It was merely my feeling that the norms of efficiency should take their cue from human beings, not from machines or abstract models conceived by narrow-thinking engineers. In fact, it seemed to me that the inversion of the relationship between human and machine was bad partly because it was *in*efficient—it subjected people to outlandish inconveniences and indignities as they struggled to meet the needs of pieces of equipment.

While threshing with the crew, I had noticed something interesting. The work was heavy and the day long, yes. But there was something pleasantly haphazard about the scheduling; there were lulls. Lulls waiting between wagonloads. Lulls caused by lack of coordination of the persons overseeing, if anyone was overseeing. Lulls for eating and drinking. Lulls here, lulls there. If I hadn't been alert to the question, these gaps could easily have been overlooked. The lulls did not constitute mere empty time; conversation, for instance, often continued unabated when the work stopped. Lulls were part of the natural flow of human activity and rhythm. They were a testimony to genuine human leisure.

So far my conclusions about Minimite successes had been based on my own experience and subjective impressions. Yet while Mary and I were gaining in proficiency, we could not speak for the Minimites themselves. For one thing, we still did not begin to approximate the time- and labor-savings, born of experience and skill, that they manifestly possessed. On the other hand, we didn't farm nearly as much land as they did. And the Minimites, taken individually, had different work habits. Some liked to take life easy; others applied themselves punctiliously to the day's work. Just because an opportunity for leisure existed didn't mean everyone took advantage of it.

The timing of threshing could provide not only greater objectivity, but universality. The job was one that both the Minimites and I had performed—and the work came as a package, averaging together the efforts of different laborers from different farms. What's more, it was avowed to be among the most strenuous. Thus it gave a sense of the maximum Minimite workload, the peak at the opposite end of the trough of lighter intermittent work in the off-season.

On two different afternoons during the threshing harvest, using a digital watch, I set out to clock how I spent my time. I used results from a typical workday at James Stoltzfus's, another brother of Gideon, Elbert, and Alvin. Some strands of conversation are included.

12:30 P.M. Left home. Walked to James's. Waited as the crew assembled.

1:13* (*Time interpolated). Rode on wagon down circuitous track to field.

1:22. Threshing begins.

1:37. Five minute break waiting for empty wagon.

1:42. Second load.

2:02. Third load. The new wagon was waiting for us, but we slowed down as crew members dropped their work in astonishment while I was telling them about life in Boston.

"The main industry in Detroit is cars; in Pittsburgh, steel; in Boston, education. There are one hundred thousand students in Boston at fifty different colleges."

"What percentage actually do what they learn?" Elbert asked after a pause. The answer was very few, because if they don't change their minds by the time they graduate, they often discover that what they were learning has already been outmoded.

2:27. Fourth load.

2:46. Fifth load and break. Now we were talking about New York. In New York, I explained (though I'd never lived there), everyone had to be paid more money because it didn't go nearly so far as it would anywhere else.

"How much for a regular house?" someone asked.

"In the city? I have a friend whose dad lives in a three- or four-story brownstone—a skinny house all packed in next to other ones. Probably cost a million dollars." (It was down the block from the Guggenheim.)

"What about just a regular apartment?"

"That depends on where it is and how many bedrooms and how nice."

"Just a regular apartment."

"Three bedrooms? Three bedrooms cost"—I was venturing a wild guess, based on my Boston experience)—about twenty-five hundred dollars a month."

"Aw. What about for just one person?"

"A thousand dollars. That would be on the cheap side."

"You must have to be rich people to live there."

"A lot of them are lawyers. My friend's dad composed musical scores for television shows and conducted Broadway musicals."

3:26. Sixth load.

3:41. Seventh load.

4:02. Eighth load.

Elbert had heard a rumor about me, and began to follow it up. "Parents divorced?" he asked.

"Yeah," I answered.

"How does that make you feel?"

I didn't really know and couldn't say. Where feeling should have been was just an empty spot.

4:20. Ninth load.

Elbert said it would give him the jitters to live in the big city. There's "no space."

4:41. The supper bell. I shinnied up the wagon standard, got a good seat on the straw, and as the horses moved, began to sway back and forth like a pampered swami upon a divan. Up high there was a good breeze. (In the hollow where we had just been, there had been next to nothing.)

4:50. Back at barn. Waited as everyone assembled.

5:20. Supper.

6:00. Tenth load.

6:20. Eleventh load.

6:41. Ride back.

6:50. Unloaded last sheaves into the threshing machine.

7:15. Walked home, stopping at the Minimite general store to buy a straw hat.

8:20. Retired for the evening.

Thus ended my time-motion study. How did it all add up? (I performed the same exercise twice and got almost the same findings.) I had expected some mitigation of the day's demands, but nothing had prepared me for this: while seven hours and fifty minutes were available for the afternoon's work (mainly threshing), only four hours and forty minutes had been spent in actual physical work. One hour and fifty-seven minutes had been devoted to breaks (including two "accidentals" and one meal). An additional one hour and thirteen minutes elapsed walking to and from work, shopping at the store, and waiting for the crew to assemble at the beginning of the session. All this time was sheer cushioning—the benefits of camaraderie, conversation, fresh air, and natural scenery, without the labor.

This was a half-day's work. To get a full day's approximation, the four hours and forty minutes would have had to have been multiplied by two. If this were done, the equation would have yielded nine hours and twenty minutes net labor.

Admittedly I had been spared some equipment setup time, which I spent waiting under a tree. I was also unable to factor in time for daily chores, which could vary considerably from farmer to farmer. But even if all this had been added in, say totaling an hour, it may have been partly offset by lulls that usually occurred in the first half of the day: the extra meal and the ten to twenty minutes of daily devotions after breakfast. Lunch was considered the big meal, and this was omitted from the afternoon's calculations.

Nine hours and twenty minutes actual labor in the peak season. Beneath the huffing and puffing I began to discern the outlines of a tolerable busy-season workload. My study revealed nothing of the free time during the long months of the off-season. The Minimite farm was no modern-day factory, where the work proceeded relentlessly year-round. Much less was it a New York law firm, where the typical partner spent sixty to eighty hours in mental millwork per week.

The threshing season lasted approximately two to four weeks.

But were these efforts sufficiently productive? Considering how leisurely the work was and how minimal the timesaving equipment compared to that of other farmers, was the ratio of output to input efficient enough for the Minimite farmer to survive? I collected no hard data on the financial status of the men I worked with, but I could make the broad observation that while their agribusiness brothers were dropping right and left from competitive pressures, not a single farmer in the community had gone out of business in the fifteen years since the settlement was established. Taken as a group, the community was debt-free. This may come as no great surprise when one realizes that the largest cost of farming nowadays and the likeliest reason for bankruptcy—in short, the heaviest drain on the farmer's time—is the timesaving equipment itself.

—

The financial success of the community was the more remarkable considering that threshing didn't even pay. Thanks to combines, wheat prices were too low for the Minimites to compete. They simply ground up the wheat in feed for the horses and other farm animals and saved the straw for bedding. We worked those daily nine or so hours for the comfort of quadrupeds.

On reflection, there was something odd going on here. For if one cared to notice, the greatest demand for horses in the community was created by threshing. The need appeared circular. Threshing was needed to feed horses so that they could help thresh. Horses, in turn, took more land for grazing, hence horsepower to pay for it. Words came back to me from the local sage, Cornelius, who pooh-poohed the notion of horse-ownership at all: "They're just turning the machine." The question was Who? The horses or the humans?

Had the Minimites, unwittingly, become the servants of their own technology—namely the threshing machine—in the ultimate service of voracious horse appetites? Horses, in this view, can be taken as a kind of technology themselves, an organic engine occupying the top of the food chain. Was the god the Minimites served Judeo-Christian or equine?

Or was it possible that, whether consciously or unconsciously, they clung to the work for its intrinsic benefits—as a sort of recreation?

Either way, the upshot of this speculation was incredible: threshing, the most onerous of Minimite tasks, was discretionary. Could I be wrong about this?

The final day of threshing for the season proved propitious for raising the question. The farm was Wilbur's, the thoughtful fellow in line for the ministry. And because rain had interrupted the harvest, the sheaves were soggy and heavy. Into a small group of groaning threshermen I plunged—Wilbur, Elbert, and Bill—hoping to make my case.

After a few minutes of doleful heaving, Elbert muttered, "How many acres do you think this field is?"

Nobody answered. Cars were driving by one after the other, stirring up a cloud of dust from the gravel road. We coughed and batted at the air with our hands. Wilbur finally turned to me and smiled as if to concede victory to me and my ilk. "Do you think we'll make it?"

I let down my fork. "Of course!" I replied. "You're debt-free within your group." Most motorists here, ironically, came in search of Minimite vegetables and baked goods sold at roadside stands throughout the settlement. "These people are making you money. That is how you survive."

"I guess." The response was barely audible.

In the ensuing silence, I cut to the chase. "You are doing most of this work just for the sake of horses," I announced. I knew if he was like his brother (who I had learned was hardworking Howie), he had six to ten. "What if you just got rid of them?"

He looked at me quizzically.

"Now you need to pay for land, harnesses, bridles, all the equipment, feed, and of course the barns—not to mention all the work that goes along with it. How much land would be enough to feed a family? This field?"

A light went on in his head. After a moment's consideration, he said, "You'd better believe it. It'd have to be an awfully big family."

"Then what if you just did it—got rid of the horses?"

"How'd you turn over the land?" he asked.

I realize I'd caught his interest. Now I was reeling him in.

"Shovel." I had done a little reading. Not only could a spade eliminate the cost of horses, it also would make it possible to preserve the structure of the soil. "The structure is in the layers. The soil is not just dirt thrown together. It's crisscrossed with worm tunnels; it shifts on its own with the frost. If it could be allowed to develop naturally, there's no telling how much productivity could be increased. Look. I don't know if this is true. It's just something I read."

Wilbur thrust his fork down, then leaned back and considered the idea. "I could see a way it'd be more productive. If you turned the soil over by hand, you'd be very careful not to waste one spot of land, and in that way, you'd be more productive. That's the only way I'd see it. The rest of what you say, I don't see." But he asked, "How'd you get places?"

"With less land, you'd all be closer together. You could walk. They say that's the best form of exercise. You could even ride bikes. That's one of the best ways to exercise your legs."

"Get your heart pumping," threw in Bill.

"But," Wilbur returned, "You'd need money. To pay taxes, buy land—"

"—medical costs. Yes, you could have some way of making a little money to pay for those."

"Yeah, what you say so far, you could do it. I can't refute your theory. You'd need two acres for produce, one for timber; and your flour—you'd need another acre for wheat. And one for the dairy cow."

As discussion deepened, the weather began to change. Thunderheads were gathering as if from every direction, bringing muffled booms. It grew dark.

"The main problem with your idea," Wilbur continued, "is, well . . . I can see you wouldn't have enough—to fill up on. Six acres. That'll work you a while, but—"

"You mean you'd run out of work?!" I clucked. He had conceded the point. Yet his phraseology was revealing. Work was a kind of nutriment. Something meeting the body's needs and satisfying the soul.

"See Bob Esch's over there? He has ten children. They work eighty acres, well, fifteen are in pasture, so sixty-five acres. Now, with that many children growing up, and you've got every weed pulled, and they say 'What'll we do next?' And you say 'Find more weeds to pull' and they ask 'Why?' 'To keep busy.' You can keep that up only so long. We're not strong on working away."

Dark rain clouds closed in against the trees of the neighboring woods. Thunder began to crack. But a small patch of blue sky still gazed down on us. The circle we occupied below remained, almost miraculously, dry and tranquil. We threw up the last few sheaves to finish the load, and we shuffled over next to the woods and sat down. We continued to talk.

"Well," I said, "you mean that without enough work to do at home, they'd have to work outside for someone else instead?"

"Yes."

"So then you would want to make work for yourself, to fill up the time?"

Wilbur chuckled skittishly.

"But that's exactly where I'd come in. For me, the kind of life I'd like to have, I'd work hard half the day and read and study the other half."

Wilbur sighed with recognition; it was the sort of thing he would have expected a college-educated person like me to say.

"Just think how much of the Bible you could learn," I cajoled him. "Think how many good books you could read."

"But here's a place where we might *disagree*"—on this word

Wilbur's eyes popped out in gentle remonstration—"with your way of thinking. We don't go for book learnin'." The sequence of logic that followed, nonetheless, fell out with textbook precision. He knew his destination. He had several planks on hand. And he built his way, plank by plank, until he had arrived. He called into question "intensive study." He repeated the phrase several times like a physician reminding a patient of the dangers of high cholesterol. He painted a verbal picture of a bifurcated personality—one whose head swam with ideas that were out of sync with his daily reality. By contrast, "to live the way we do, you have just what you need to know. And you learn that—in *action*." After all his careful preparation, he landed on the last word with zest.

I couldn't have said it better myself. So I took a breath and tried a new flank. "It's one thing to work because you have to. But you're toiling in order to gain luxuries you could easily do without: the fancy foods, the buggies. And most of all the horses. A horse would be considered a luxury in most parts of the world. In a peasant society, only the rich own horses."

"We aren't greedy," said Wilbur. "It's just that other people are doing it." He seemed to ponder the thought. "They weren't satisfied . . . and thought more would satisfy." He looked troubled.

Then I told the tale of the southern fisherman: "The rich man from the North came by one day and saw the southern fisherman sitting, just sitting by the water. This horrified him. 'What are you doing?' he asked. 'I'm sitting,' replied the fisherman. 'Why aren't you out there fishing?' 'I have caught enough fish for one day,' he said. 'Don't you know,' returned the rich man, 'that if you continued, you could earn more money, and with that, buy another boat? With two boats you could earn more money still and buy better nets. Then you could catch even more fish and pretty soon you'd have a whole fleet of boats. Then you'd be rich like me.' 'What would I do then?' asked the fisherman. 'Then you could really enjoy life.' Replied the fisherman, 'What do you think I'm doing right now?'"

Thunder pealed out from behind the trees, and those ominous black walls seemed to be collapsing on top of us.

Elbert, who had delivered the last load to the barn while we chatted, reappeared with a fresh team. "If Eric talks anymore," Wilbur called to him, "we may be doing a lot less of this next year."

"What?"

"He's persuading me to scale down our operations."

"Well, let me know, so I'm not the last to go small."

I repeated for Elbert the tale of the southern fisherman. He snorted and laughed and said, "Why, that about says it all!"

But Wilbur's eyes darkened mischievously, and he shook his head at me. "An idle mind is the devil's workshop!" he declaimed.

I took the words as a compliment. Whatever reservations Wilbur still held, he had conceded my main point: whether glumly or gladly, he and his fellows labored under a surfeit of horsepower.

In a few minutes, the storm blew over. It never did rain on our party.

Wilbur operated a small harness shop, housed in a low, gray building of metal siding next to his barn. It was open for business to Minimite and non-Minimite alike. I dropped in one day to take up the conversation again, entering a dim room lined with dangling leather straps, chains, halters, and bits. Nobody seemed home. "Wilbur?" I asked doubtfully. There was no answer. But I noticed through a doorway, back in shadows, the silhouette of a seated figure. "Wilbur?" I took a few steps toward the doorway and heard the person say:

"What have you been doing today?"

I started. It was Wilbur, but his voice seemed strangely lackadaisical, almost sing-songy. I couldn't see his face in the shadows.

"Picking tomatoes."

"There's a lot to do on a farm," he replied in the same sing-songy voice. I sensed he wasn't in the mood to talk and left.

A few minutes later I happened to bump into Bill, Edward's farmhand, and he was breathless with news. Edward had visited Wilbur just before I did. And there had been an incident. Bill had a way of reporting facts selectively that piqued one's curiosity: "Wilbur was sewing on the sewing machine, and it got hung up, and Edward pointed his finger and said, 'There's your problem.' Wilbur ran the needle right through his finger."

My jaw fell. "Why did he do that?"

The machine, apparently, had stalled while the two had their tête-à-tête. Then Edward looked, saw what the problem was, and pointed. And Wilbur ran the needle. "Wilbur claims it was an accident."

"Why would anyone stick his finger right under a sewing machine needle?"

"There's Edward for you," said his loyal hand. "Sticking his finger in other people's business."

Two of Edward's other fingers, by the way, were truncated. He had stuck them in a meat grinder just as someone turned the handle.

Could a man of such gentle demeanor have had it in him? That might have explained his strange behavior. Wilbur never could bring himself to agree with me in words. It was only through action that he made the point! I couldn't help chuckling. Perhaps Wilbur, at last, had given the devil his due.

But if so, it was only a flesh wound. The main battle was still to come.

harvesting

as the pumpkins turn

As the residue of James's sheaves was cleared away, the threshing season breathed its last. A new, cleaner spirit seemed to aerate the land; the days were getting shorter; the sky at night was clear and crisp and starry. Fall was coming.

One bright morning I looked up from the notes I had been accumulating from all the recent happenings, and there stood my wife, as in a vision.

"It's as if you're not even here," she rebuked me.

I smiled. I noticed a dimple between her eyebrows—maybe you could call it a crease—and for the first time in I don't know how long, it registered.

"I want you back," she said, and the creases around her eyes deepened into crow's-feet. She was, I inferred, commenting on the long hours of journal keeping I had logged since the heat wave had broken. "I'm beginning to wonder if you married me just so you could write your book."

The jab hit its mark. As her comment warned, there was a way that reporting on the events—if I weren't careful—might intrude upon those very events. Heisenberg's assertion—that measurement can change the nature of the measured—was still in force.

But almost before I could lift a pen, events interceded.

•   •   •

On the last Sunday of August, as the Minimite church service let out, an older man I didn't know too well approached me. He apparently knew me, however, and had been keeping tabs on my pumpkin patch, which was visible from the road. Now he said with a shake of the head, "Looks like your pumpkins are dying off pretty fast. At least they'll be good for selling some small ornamental types. I don't blame you 'tall. We grew a few for ourselves, sprayed every week, and couldn't keep the blight off."

What? The pumpkins weren't supposed to ripen until around the first of October, the opening of the Halloween market! I rushed over to the field. Under the curling leaves, mature orange pumpkins were indeed visible. I inspected them more closely. For once the Minimite diagnosis was incorrect. The leaves were dying off, yes. But the pumpkins were jack-o'-lantern size. Perhaps the rains of July and August had hastened their growth. I was transfixed by the many exotic shapes—long and wienerlike, short and squat, round and fat, some a little twisty and gourd-shaped, others pointy, and one a twin pumpkin—two merged into one. Most were a rich orange, but many were still half green and some were checkered. The average size was somewhat smaller than the seed company had said to expect, but this I attributed to the fact that I had not thinned them out as well as I could have. What we lost in size, we made up for in quantity.

But how would they keep?

We at least had to get them out of the sun. The next day the Miller boys helped us fill up two wagonloads, but that was all the time they could spare. The sun was sinking, but we knew it would rise again in a few hours, and we were on our own. We still had half a field to pick! As we fretted about what to do, a big, orange, opalescent disk slipped over the horizon, like an answer. It was as if someone had overheard our discussion, tripped a switch, and turned on the auxiliary lighting system. The harvest moon had risen.

We were unaccustomed to working after sunset, but we took the hint, returned with the Escort, and began to fill the hatchback. The moon soon rose higher and became more luminous. We were amazed at its strength. We returned again and again to the field collecting pumpkins in its soft radiance, awed by the romanticism of the escapade. But beauty was married to function, and the blue of the light, the soft night breeze, the myriad eerie silhouettes, and the suspense of the final outcome in no way detracted from our efficiency.

It was here in the pumpkin patch, I suspect, when we least were thinking about it, that Mary and I at last consummated our marital pledge of mutual surrender. No, more than this. We entered into a fullness of being with roots reaching deep into the earth, into culture—a clasping unity elevated far above anything we had yet known, a state of ecstasy from which there could be no looking back. It was as if the field were there to harvest us, not we it, the whole undertaking a pretext, a cosmic matchmaker's ruse. At the stroke of midnight we shed our mortal shells and became prince and princess of creation, presiding over the majestic ball of life, ceremoniously joined with nature in jocund betrothal, a feast of love.

But the practical output of this orgy was eleven hundred pumpkins. The next morning our awakening was bracing.

We tried filling the front porch first, and when that was done, moved to the bedroom, and from there to the backyard under a big tree. We were reluctant to store them outside, however, because unless they can be kept dry, pumpkins will rot quickly at the stem.

We put up a sign marked "PUMPKINS" over our mailbox. After selling only a handful over two weeks, we put up another sign where our road met a state highway six miles away.

Pumpkins were one of two cash crops we had planted; the other was sorghum. Now that Sylvan had provided ground for our sorghum, we were responsible for helping him with the cooking at harvest. (Whenever I came by to work on my crops, Sylvan still arched his dark brows in his usual expression of mock alarm at the brashness of an unskilled beginner.) The day was approaching for the appointed collaboration, and I dearly hoped I could be of some use. We stripped the tall sorghum stalks of their leaves, then cut them off near the ground using machetes and piled them on wagons.

—

"Ouch!"

Not again!

I didn't know which the collision hurt more—my toe or the pumpkin. Either way, it hurt me in the long run. I tried not to think about it. The trouble was, it was difficult to walk to the bathroom in the dark with so many pumpkins on the floor. Several had already gone soft around the stem and begun to blacken after only a couple of weeks. And now I was hastening the deterioration by chipping their skins with my big toe. They rotted through such openings. Muttering to myself, I leaned over and picked out bits of orange rind from under the nail, and then went back to my journal.

—

Before sorghum juice could be cooked down, it had to be squeezed from the stalk. And since it soured in a few hours, squeezing took place the night before it was to be cooked.

When I arrived at Sylvan's for the long-awaited occasion, the house was empty. I sat on the porch and waited until the moon had risen. Blank objects loomed, pale and ashen, here and there in the lunar glow—a barn, a fence, a tree—and behind, a darker row of woods. The scene was downright spooky. Eventually I got skittish. Could he have forgotten? We had set the date over a week ago.

Clomp-cht-cht, clomp-cht-cht, clomp-cht-cht. . .

A horse was approaching. The scrape of buggy wheels rounding a corner rustled on my eardrum. A fuzzy shape grew in the distance. A lantern light appeared. A jolly baritone voice crooned out of the blackness, "Well, if it isn't our visitor." Then I saw them: Sylvan, Ida, and their baby, suddenly lit up by moonbeams and gazing out at me from the bench of a spring wagon. Sylvan panned me with his usual mischievous grin.

"Boy, I can see you so clearly."

"Harvest moon," he announced. "Good for staying out and working at night."

Sylvan had pulled up to the side of the porch, and his wife, nodding politely, carried the baby into the house.

Then we went to work. The sorghum press looked like the wooden frame of an unfinished teepee. It was a simple device made of a tall, central rotary drive shaft with a diagonal member supporting a single crossbeam, to which the horse (which Sylvan was now leading towards it) was hooked. Under the shaft was a narrow set of metal rollers that turned when the horse went round. The whole contraption sat in an open area near the barn.

Sylvan began to insert the sorghum stalks between the rollers, and as they flattened, juice trickled into a pan that emptied into a tube that, in turn, drained into a large holding tank down the hill. On into the night we worked at it, with me leading the horse and Sylvan inserting the sorghum over and over and over. I made it through without a hitch.

At morning's light, Mary and Ida joined us. Sylvan tripped the cock that released the juice from the big stainless steel vat. The sorghum trickled into the first compartment of the long, flat sorghum pan, following its mazelike divisions through a series of hairpin turns. Meanwhile I sawed wooden pallets into pieces small enough to throw on the fire that was roaring under the pan. All day long, under the brilliant blue sky in the cool fall air, with leaves spiraling down slowly in the stillness, the syrup bubbled and thickened and wound its way through the compartments until it reached the final holding section. Here Sylvan measured its temperature with a candy thermometer, and at the precise degree, let a portion drain into a small jug. When the jug was full, Ida replaced it with another, then another. Periodically Sylvan skimmed off greenish foam that rose to the surface and retrieved drowned yellow jackets with a dipper. I continued to cut and throw pallets while Ida continued to bottle the finished sorghum and Mary affixed the labels.

By day's end, my wife and I were proud owners of several hundred jars of molasses with a retail value of about a thousand dollars.

Cooler autumn weather not only made for pleasant sorghum cooking, it also had one other welcome effect: it refrigerated our pumpkins. By the first day of October, over ninety percent were still intact. Customers began streaming in, and by the middle of the month almost no pumpkins remained. We sold about 400 at the full retail price of fifteen cents a pound, and 450 wholesale at seven cents a pound. To help us out, the Millers sold another hundred from their own stand at retail price.

The pumpkins netted about a thousand dollars, which, together with the prospective profits of the sorghum (still to be sold), put us in very good stead. By these means, and with a little extra work I had begun to do for Mr. Miller to trade for rent, we were able to grasp onto the goal that at first had seemed so daunting: to earn our living by the labors of our own hands.

—

As nature set the timetable for a single day, so did it, conveniently, for the whole year. Late fall provided Minimites leisure and cool weather for festive events. Barns would be built, new families would move in, other families would move out, young couples would marry, and old couples would take up residence in the *Grossdaddi Haus*. And there was still more spare time—time to take stock and reflect.

With cold weather came a pleasant hibernation. The first snow flurries, seen through the window on a day when we sat by the wood stove writing letters, filled us with indescribable serenity. Our cozy haven seemed so much cozier.

The season was not without subsistence activities, but they lacked the push of warm-weather work. They tended to be tasks that could not be performed at any other time or were not pressing. The pig we had been feeding through summer and fall, for instance, now approached his appointed destiny. We could slaughter and prepare the meat in the chill out-of-doors without threat of spoilage.

Another job saved for cold weather was fencing, and Mr. Miller allowed me to work off our rent by locating and cutting cedars from his woods. Cedars make good posts because the resin that permeates their trunks acts as a natural preservative. Seeking and cutting these sleek trees deep in the forest while inhaling the scent of fresh-cut cedar wood was patently satisfying.

• • •

And there were other off-season duties as well.

"One, two, three, four, five. Now release."

Mary, who was flat on her back on the floor, relaxed her pelvis and her huge belly rose gently up again, like a bowl of bread dough.

"Again," I said. "One, two, three, four, five. Release." After sinking in a second time, the engorged abdominal region rose again and swayed. The exercise, described in the Lamaze manual we were using, was designed to stretch and flex Mary's lower back, which had become stiff and sore from all the extra hoisting. It worked.

"I can't believe how much better that feels," Mary sighed after a few repetitions.

I leaned over to the bedside table for the Lamaze text, and my eye fell on the part that insists on the husband's continuing marital relations as far as possible into pregnancy for the sake of the wife's self-esteem. I looked at her from the side, taking in the satiny curve of her cheek. I was never one to ignore instructions. Another attempt to complete the exercises kaput.

The next day, for a change, we made it all the way to the breathing exercises.

"*Whoo*—whoo—whoo—whoo—whoo—*Whoo*—whoo—whoo—whoo—whoo—" The rhythmic panting suddenly broke off.

"Wait! Forty-five seconds to go!" I urged her, rattling my electronic wristwatch in the air. But I saw she had given up entirely. She had lost her train of thought. "Try again."

"*Whoo*—whoo—whoo—whoo—whoo—*Whoo*—whoo—whoo—whoo—whoo—"

We persisted in the pattern this time until the full ninety-second mock-contraction was over. Mary gasped for air, jubilant, eyes aglow.

"I think I'm getting the hang of it."

The session ended even more desultorily than the day before.

We still had to master the pushing exercises. For a day or two, we forgot about them entirely, then suddenly as we finished the dishes, it struck us.

"*Push*. Two. Three. Four—Now remember to relax the pelvic floor. Eight. Nine. Ten. *Push* and relax at the same time. Twelve.

Thirteen. Take a quick break. Release a little. Now PUSH again. Two. Three. Four . . ."

Pushing was, if anything, easier than the other exercises. The only trick was for Mary to learn not to push too soon in delivery, or the emerging baby might impact the not-fully-dilated cervix. She had to learn also to *un*-push. "Okay, let out all your breath and take five or six short breaths. Now you have nothing to push with."

It was the best workout yet, so I rewarded Mary with a massage.

"*Ooooo*. That feels *won*derful. Some more there. Yes."

I rubbed the rest of the pain away, and soon enough we turned over and went to . . . bed.

SEVENTEEN

after the fall: even farmers get the blues

I just felt so sorry for you sitting there so lonesome, I had to come over and buy one," said the man dressed in Western gear who had swaggered across the street from the Acme Boot Outlet. A gust of wind howled past, and this time, thankfully, my SORGHUM MOLASSES/ SWEET POTATOES sign did not blow over.

I smiled and, removing a glove, pointed to the jars. "This one's two-fifty, this one's three-fifty, and this one's. . ."

"Oh, just give me anything," the man interrupted, turning to look at the blond woman who, I supposed, was waiting for him. She was seated in the passenger side of a yellow Corvette in the Acme parking lot, and returned a languorous smile.

"I recommend this," I said, and he handed me three dollars and fifty cents for the medium-size jar. The label read:

Old-Timey Cooked
SORGHUM MOLASSES
(Cooked over a wood-fired pan)

As the man walked away, it dawned on me that perhaps I did look a little lonesome and forlorn. He was all set to hop into a souped-up vessel of comfort and control, but I had no visible means of transportation (my car was parked around the corner of the building). He must have assumed I was one of those local "Amish," eking out a bare subsistence, abjectly dependent upon the patronage of caring folks like himself for my evening meal.

It was a blustery December day and I was seated at a table laden with sorghum jars, on a traffic island catty-corner from the regional shopping mall. I was reading Albert Camus' *The Stranger.* And I was wearing a wide-brimmed straw hat.

It was all so convincing, except for the existentialism.

I almost said something, but a scruple forestalled me. Best to let him go on thinking it. It would take a while to explain, and judging from his current preoccupation, I doubted he would even listen. Besides, today for some reason, I was starting to feel a little like the character in *The Stranger.* Maybe the fatigue of long anticipation was catching up with me, a kind of male counterpart to morning sickness. We were now in the third trimester, and time seemed to have slowed down to a crawl.

At the end of one long day, I went to Sylvan's to see if he could sell my leftover sorghum jars to wholesalers for me. The path to his farm was well traveled. Sylvan's place was bustling with straw-hatted workers, rat-a-tat-tatting and sawing, swarming around an almost-finished pig house. The men paid me almost no heed at first. Then Sylvan looked up, arched his brows, smiled, and said, "Even unto the eleventh hour, the worker is still good for his wage."

Low sorghum sales all day and now this—a biblical exhortation to get me to help finish the farrowing house. I don't know why I was so reluctant to do so. I simply wanted to get my money and get going. Maybe I'd spent too much time lately behind the sales table, listening to the sound of jangling coins. The price Sylvan quoted was a ridiculous twenty dollars a dozen (I was getting forty-four selling it myself), but that was better than nothing.

But before I even had a chance to clear my throat, Sylvan's brother cracked, "Does that mean you're gonna feed him dinner like you did us?"

With the snickers that ensued, I worried if I was being pegged. Although my ilk stopped burning them at the stake quite some time earlier, some of them still couldn't quit reminding me. Consoling myself with the thought that they were nearly finished anyway and that, given my inexperience, I would probably not have been good for my wage, I skulked back to my car, feeling like the *Stranger* I was.

Sometimes I will admit a stray joke or sidelong glance hit me the wrong way, with a force that was almost palpable. I suppose my feelings of injury revealed the development of an underlying allegiance. Could it be that for the first time in my life, I had begun to feel as if I really belonged? To have this suddenly pulled out from under me, or even to hint at its being pulled away, made me disconsolate.

The silences of Jed, one of Mr. Miller's older sons, had become unnerving. To receive one of his recriminating looks was to be indicted by a single-member grand jury. He thought I was a freeloader. Was he tired of helping me out as one of his dad's cheerful assistants? He made very sure that I didn't take too much kerosene for my money (impossible to do since the two-and-half-gallon capacity of the jug was the amount I paid for).

Admittedly there was some evidence for his case. Two days in a row I had made him wait when he came to help me, and I found him sullenly lugging stalks of sorghum and leading the horse by himself. And before that, when he and his brothers were trying to locate the buried pipe from the spring to our house, Amos had had to rouse me from my afternoon nap. I arrived without jumping in to dig because it took some time to come out of the fog. Sleep makes me dropsical. After thirty seconds one of the brothers turned and, pointedly, put a shovel in my hand.

Didn't they understand the sacredness of naptime?

Even Mary, now and then, seemed to lose faith in me. I left a plastic bag containing ten pounds of frozen venison an "English" neighbor had given us out on the stoop overnight to keep it cold, as I had done on several other occasions. The next morning she asked me, "Where did you put the deer meat?"

"By the back door."

"I don't see it."

"It's right there. Right on that cement block. Don't you see it?"

"There's nothing there."

I looked again. There was nothing there. Not a trace of the package. Not a drop of blood. Not a shred of plastic. No paw prints. No lingering dog breath. Nothing.

Worse than the loss of food was the loss of face. I thought I was invincible. I set up conditions for the perfect crime—and it occurred. (The assailant, I knew, lived beyond the long arm of the law, and was now probably dining in luxury in his canine pad next door.)

Mary looked at me with a pout of unrequited hunger.

"I'm sorry," I pleaded. "I'll never do that again. But it worked the other times."

"I told you that it was dripping on the outside."

"You did? I knew you were saying something but I wasn't really listening." Oops. Shouldn't have said that.

Back to the stored chicken meat.

Our canned chicken, however, could easily be confused with our canned pig. This was because we slaughtered and preserved our chickens and pigs on successive days and the 150-odd jars in which we placed the meat snuggled amongst each other in one giant arc surrounding the pressure canner. Was it chicken or was it pork? The fat for the chicken (which rises to the top) is yellower. But the numbers came out wrong when I tried to divide jars by fat color. We didn't really know what meat we were eating until we bit into it. Even then it was hard to tell.

One day Mary served sautéed pork as the main course of a meal accompanied by sweet potatoes and apple crisp. The chunks of pork were white, tender, and well seasoned.

"This almost tastes like chicken!" I exclaimed. Then I paused. Mary frowned. Was it chickenlike pork or porklike chicken? We didn't know.

But perhaps it was best not to think back to the preparation of the meat. As we prepared to behead a live chicken, Mary would hold the bird's body and wings, looking away while I stretched out the neck and lowered the axe. The first stroke was usually lethal enough, but since I was fearful of striking my own thumb, several more were necessary to sever the last connections. One time, however, I half-missed on the first stroke. The chicken sailed out of Mary's arms with blood spurting everywhere. I don't know why Mary chose to wear a chiffon blouse for this occasion. It took a while to catch the thing and finish the job, and it was the end of the blouse.

Our pot of boiling water for feather-plucking proved to be too small. The first time I dunked a fresh-killed chicken into it, the water overflowed all over the stove, creating a gooey and unappetizing slime. So I poured out some of the water. This left us too little. The remaining bath did not cover the dead chicken, and we didn't have any more hot water on hand. I tried ladling water over the chicken in hopes that this would loosen the feathers. Mary, meanwhile, was becoming demoralized. Rivulets of lukewarm bloody-brown liquid coursing over the carcass of a zonked chicken emitting a chicken-death-smell made her gag. As the water cooled, rigor mortis set in. It had been difficult enough battling a chicken when it was alive and wanted to stay that way. Now that it was dead, its feathers were still hanging on.

—

As a musician I tended to tune in too closely to my surroundings. When there was dissonance in the air, there was dissonance in my psyche. I could even internalize changes in the weather. This made me especially susceptible in the wintertime and might have accounted for some of my sullenness. When the sky darkened, so did my mien. It was not the long winter evenings by kerosene lamplight that got to me; it was the short gray days, the low leaden cloud banks. The very heavens seemed to brood. Because of a long, low porch on the south side of our house, little light penetrated our living quarters even when the sun shone. A standard sixty-watt incandescent lightbulb might have helped.

Lacking this, I began to feel that playing some cards would cheer me up—specifically a good game of bridge. The thought hounded my light-asphyxiated brain like the craving of a heroin junkie for a needle. My parents had belonged to two bridge clubs, and growing up, I had sometimes filled in for absent players.

A retired couple we'd met once in town had mentioned that they played and gave me their phone number in case we ever wanted to hook up. We didn't have a phone, but that didn't stop me. "Mary, let's go for a ride."

"What, right now? Where?"

"To the Brailowskis'. To play bridge."

"Who are they?"

"Don't you remember meeting them in town? They said they'd like to play sometime."

"But I don't know how to play bridge."

"That's all right. You can be the dummy."

"Thanks a lot. So we're like just gonna drive over to these people we don't even know and announce the bridge game to them?"

"We can call them from the Minute Mart."

"This sounds a little . . ."

"Pretty please . . ."

It was twenty-five or more miles along narrow roller-coaster roads to the opposite end of the county and a small lake I had never seen before. Mr. Brailowski greeted me at the door and gaped as I rapidly repeated my gratitude for having us over. During the game I had the feeling he and his wife were exchanging sidelong glances. To keep Mary in the dummy position, I drastically overbid my hands and dominated the play. Our score was abominable. When we finished for the night, the blank looks of the Brailowskis reminded me of the blank cast of the sky. I was worse off than I had been before.

———

I shouldn't have turned down the work Sylvan had offered on the pig house—such inspiriting activity might have been just what the doctor ordered. Instead, now Mr. Miller shifted me from cutting cedars to digging a drainpipe trench in a cow pen oozing with recent defecation after a rain.

The first time I went over for the project, Caleb was feeding calves a short distance away. I heard a din that sounded as if a building had come crashing down, and looked up. The calves were all crowding around a bucket of grain that Caleb had placed in their midst, and one of them must have kicked it. After I'd been digging for a few minutes, he sauntered over, and I accosted him.

"I know you're jealous of me having all this fun here."

He paused and smiled. "Can't say that I really am."

"Well, if you ever get jealous enough, you'll be more than welcome to come and take over some of this."

He smiled and edged a little closer.

"I can't believe how little I've got done." I glanced over my shoulder at the measly ten-foot-long-trench that seemed to be filling back in as soon as I dug it.

"How long've you been at it?"

I looked at my watch. "Oh, about an hour."

"What do you think's making it go so hard?"

Oh, oh, I thought to myself. He acknowledges I'm not getting anything done. "It's not really that hard. It's just that I'm slipping and sliding. See here." I put one foot on the edge of the bank to show how slippery it was. He nodded superciliously. "Well, I guess I better go back before they eat it all." He must have been talking about the calves.

A few minutes later, he appeared again, smiling, shovel in hand.

"Oh, you didn't have to do that." I had only been joking. When I saw that he meant it, I said, "I bet you'll outperform me." But soon he was slipping and sliding just as I was. And with his slight build he couldn't get as good a grip.

"So now you're about to turn . . . How old are you, Caleb?"

"Thirteen."

"Oh, I was gonna say you're about to turn thirteen. But I guess you already did."

"Not long already I'll be turning fourteen."

"When?"

"February twenty-first."

"My, I'll have to be sure to be here that day to see you shoot up another four inches." He lowered his head shyly, and we worked a little longer in silence.

"What did you learn in school today?"

"Oh nothing, really. It was mostly review."

I asked him a few more questions as we continued shoveling until I seemed to tire the subject out.

"Do you get grades?"

"You mean do we get a report card?"

"Yes."

"Yes."

"Do they give you A, B, C, D, and F?"

"Yes. Well, A, B, C, *E*, and F."

"What does the E stand for?"

He smiled and shrugged his shoulders. "I don't rightly know. I never got one." I looked to see a huge grin on his face.

"Do you speak German in school?"

"Oh, it's about half-and-half, German and English."

Then a few more questions back and forth.

Suddenly he became more animated. "Guess you're looking forward to your new garden."

"New garden?"

"Yeah. You know the old one was so packed down, three horses could barely plow it. So it's gonna be where the orchard grass was."

I hadn't yet been advised of this development. And spring seemed a long ways away. Caleb went on to talk about the hardpan problem the mechanized farmer across from the schoolhouse was having. So compacted was his soil that water could not seep below the surface; it would just collect on top and form a lake. Everything he grew there died. First he tried disking the ground over and over again to fluff it up. But the tractor he was using was the one that had compacted the soil in the first place—it was enormous. The water still collected there. Then he tried taking a chisel plow through it, three feet down. But again he used the gigantic tractor and undid his own efforts. Finally he gave up. Two horse farmers were allowed to come in and plow. Next they planted a type of grass with a deep root system that would penetrate in one season. The ground was arable now.

"What time is it?" Caleb asked me.

"I don't know. It's too dark to read my watch." Caleb's narration had been so engrossing, night had fallen without my noticing. I'd also forgotten how miserable I was digging the ditch. I could barely see what I was doing. Still I kept shoveling. It didn't seem as though the two hours I'd committed to could have ended so soon. Caleb made no move to leave either. We'd gotten a nice rhythm going. Some unspoken bond, also, seemed to have yoked us to each other; the very muck seemed to have endeared itself to us, embracing our boots with its soft gooeyness.

We continued until the last feeble light drained from the gray dusk and I could see Caleb only as a hulking mass. It was I who finally gave in. "Well, I think it's getting too dark to see what I'm doing." He agreed, but neither of us was anxious to leave. We quit as if reluctantly, dragging our shovels back towards the house. There was a moment of uncertainty as I turned to exit. Swathed in darkness, we stood quietly a few moments gazing up at the stars. They were unusually bright tonight.

"Listen!" Caleb grabbed my arm. I couldn't hear anything out of the ordinary. "It's the train."

Yes, it was. In Pleasant Valley, thirteen miles away—so still was it this chill, starry evening. We flashed each other quick smiles and went our separate ways.

Caleb and I had never had a chance to work together like this. If I were lucky, maybe the next person to saunter by would be his disgruntled brother, Jed. Maybe that would fix something.

birth

Late one gray winter's afternoon as I walked back towards the house from the Millers', I noticed that the wind seemed to have shifted. It was coming from a more southerly direction, and I could feel its warm caresses against my cheek. I stood a moment, relishing the feeling. In the slanting light of dusk, everything had taken on an eerie cast, gray and violet, misty, with hills and rills, pools and puddles, and—what was that? I now heard a strange sound. It was soft and very high-pitched, yet a bit wild and jangly, like melodious rubber bands. It had a faint hilarity to it. I was later to ask Mr. Miller what that sound was.

"Peepers," he said. "Baby frogs."

A great cyclical turning of events, it seemed, was under way. The saga of life was rapidly being recapitulated. Tadpoles were growing legs and springing forth again from the ooze.

Amid this whole scene of percolation and restlessness, I felt new life brimming up inside of me. I could almost sense the capillary action, little threads growing together and closing the chasm that winter had opened up. The outward turning triggered an inward turning. I was coming out of my funk.

One brisk sunny day, Amos and Caleb walked into our yard leading an eighteen-hundred-pound workhorse that, they informed us, their dad had decided to lend us for the next growing season. I would be able to plow my own garden this year. He also gave consent for me to plow under the pasture behind our house so that I would not have to travel so far to plant pumpkins and sorghum. Mabel, the horse, was twelve years old, somewhat enfeebled, and surpassingly tame: perfect for breaking in a novice like me.

Under Caleb's watchful eyes, I got started. The plow skittered on the surface until it took hold. Down dove the point of the curved blade and up went a wave of grassy sod, rising, curling over, and crashing silently in a chaos of disintegrating clods. I tried to follow a straight line along the edge of the field but began to swerve left.

"To go the direction you want," Caleb said walking beside me, "Push down on that side like you would if you was steering a car." Huh? How did he know that?

I put a little pressure on the right handle of the plow and, sure enough, it righted. After I got the hang of steering again, it was like water-skiing on chocolate. I swayed and bounced lightly over small undulations of the land, with dark, loamy soil churning up beside me. To plow in rich earth like this was truly to sink one's tooth—one's single, large, iron dentoid extension—into something, an actual source of nourishment, the very embodiment of the potentially edible. It was delicious.

Mary had an intractably sunny disposition and, with the exception of that first bout of morning sickness, had never been given to major mood swings. The due date was now about four weeks away, and she continued to sew, cook, and help in the garden as if all were the same.

I, though, felt a few nervous flutters. Well-meaning friends and family members had shared their own natal experiences. In Boston over Christmas, a close male confidante of Mary's had urged her in my presence to make use of every medical screening device and procedure—and every other service on up to hospitalized delivery by an obstetrician. His own wife was due in several months, and their health insurance would pay the full cost in excess of four thousand dollars, including cesarean section if need be. In Boston, one out of three births occurred by cesarean. He couldn't believe we had let Mary's insurance lapse. The policy, though, would have cost us several hundred dollars a month and prevented our expedition. I thought the friend was a little alarmist until I learned that his brother's child had suffered permanent brain damage at birth because of a slight mishandling of the delivery.

My own father, a doctor, had asked if we were getting an ultrasound. "Why do we need one?" I replied. He explained there were some rare complications it could detect in advance, like placenta previa, the condition in which the placenta emerges before the baby. Placenta previa could well be fatal for baby and mother. He knew this because his second wife had nearly died from it, escaping only because emergency surgery was performed immediately in the hospital. Both she and the baby at one point had been pronounced dead from massive blood loss.

Finally, we learned from the Minimite midwife herself that the previous year one of her deliveries had been stillborn. Because of a prolapsed cord—the cord coming out first—the baby's air supply had been cut off. She didn't find out until too late or she would have called the doctor.

At the time I heard these admonitions, I brushed them aside. The vast majority of births went off without a hitch. I did some checking and learned that healthy mothers were as likely to develop complications *from* the hospital as they were by *not using* the hospital. Freakish conditions like placenta previa were far likelier to happen when the mother was in poor health or not taking adequate prenatal measures.

We were doing our relaxation and breathing exercises almost daily now, and I had also begun perineal massage on Mary's inner tissues to reduce the likelihood of tearing. The Lamaze book was especially reassuring. The author, Elizabeth Bing, portrayed labor as a gently, gradually intensifying series of somatic swells, possibly mild enough in the first several hours to permit a visit to the mall or the movie theater. Maternal labor sounded positively recreational. Bing also said the woman should suck lollipops rather than drink liquids as the contractions grew more intense, so we bought a bag in Mary's favorite flavor, orange. There seemed little cause to worry, but I remained slightly apprehensive.

The due date got nearer and nearer. Mary experienced the advance tremors called Braxton-Hicks contractions. But no baby. Finally the date arrived. And it passed. Two more days went by. Psychologically it was as if we had missed having a baby altogether, and we were both oddly relieved. It seemed the time would come in some distant unreal future, if ever. I no longer felt so apprehensive.

As we whiled away the days casually, rehearsing the breathing exercises, I actually was beginning to look forward to the onset of labor as a chance to break up our routine and put into practice the Lamaze methodology. I even left many last-minute details undone, confident that once labor started, there would be plenty of time to tidy the house, to gather together the various cloths, receptacles, and protective coverlets as the contractions gradually intensified. Our midwife had told us not to fetch her until the contractions were five minutes apart, and Bing had written that twenty-minute spacings were common at the beginning.

But now it was April 10th, the Easter Vigil and three days after Mary's due date. She was showing the classic watermelon configuration and the baby's head had locked in position, giving Mary a pronounced duck walk. The weather was heavenly, and in the spurt of energy it inspired, Mary worked all day inside and out. In the garden she had sat cross-legged, transplanting tomatoes. For once she overdid it slightly. Her back began to ache badly, so as the sun set, I helped her inside and gave her a massage.

Then she got a strong cramp and bent forward as if someone had punched her in the stomach. She looked up at me in disbelief. Could this be labor? The feeling was nothing like the Braxton-Hicks contractions. She described it as a wave of clenching pain that rolled up and over her midsection uncontrollably, draining her of all strength, all hope that she could possibly endure more of it. She had gone pale. She had already given up and we hadn't even started.

Before I could digest what was going on, eight minutes later it happened again, only more sharply. We were in the living room, so when the agonizing spasm passed, I helped Mary get up and undress for bed. The book said that if contractions came at night, the woman should get as much rest as possible—but the advice proved useless. The contractions were so strong, that once she lay down, Mary couldn't relax, let alone sleep.

We got out of bed and went into the living room again. When Mary sat down another cramp hit, so sharp she cried out in pain. She looked at me dazedly and said in a too-small voice, "If this is what it's like, I don't see how I can possibly make it." The tone of her voice portended disaster.

My mind began to dissociate. The room was falling into pieces. At some vague level I knew that if there was any hope, the source was me. I tried to get a hold of my wits: pregnancy, birth, what did these concepts mean? Nothing but a blank? Then I remembered: Lamaze! This was something we had trained for! I was to be the labor coach. Of course. We could begin the breathing exercises. They would help ease the pain. I had been thrown at first because we had skipped the stage with the movies and the lollipops.

As Mary sat in her chair, I gave her the cue as the next contraction started. But she didn't seem to recall how to do it. The pain was so overwhelming, it knocked her memory right out. Gently I reminded her of how it went: "Ready: *Whoo,* whoo, whoo, whoo, whoo, *Whoo,* whoo, whoo, whoo, whoo . . ." When the contraction was over, we rehearsed it a couple of times: a succession of shallow breaths, speeding up as the contraction swelled, slowing down as it abated. On the next contraction, we counterbalanced it with Lamaze breathing, and Mary said between breaths, "That's a little better."

The relief was immeasurable. Labor, which just a little bit earlier had seemed utterly unmanageable, had suddenly become manageable. I suggested we go back to bed. The minute I tried to get her up from the chair, though, Mary crumpled in another wave of crippling pain. The movement had triggered it. When I attempted to lift and carry her, this made it even worse. She was able to close the distance only by darting the few steps on her own before another full-blown contraction began.

It was now about one a.m., two hours since the onset of labor. I hadn't attempted to time the contractions yet, so I now looked at my watch. Six minutes elapsed before the next one came on. I needed to get the midwife. Just to make sure, I timed the next interval. Odd. Now Mary entered her contraction after only four minutes. Again I looked at my watch. The next one came in ninety seconds. Irrationally, I continued timing the contractions. The next was in sixty seconds. Thirty seconds. The contractions were now lasting longer than the spaces in between!

I seemed to be spiraling into a bottomless pit.

Must get midwife. Now that there seemed to be no doubt of a misreading or statistical aberration, I looked in Mary's pain-wracked face and told her the news.

But, Edna lived six miles away and had no phone.

"You can get Naomi," she said quietly. "Then Naomi can stay with me while you get Edna." Naomi . . . Naomi was Edna's apprentice and Mr. Miller's oldest daughter (who had just returned from a long visit with friends in another part of the country). She lived just over the hill. I couldn't believe Mary's brilliance.

"Come fast," Mary whispered.

As if I needed the advice. Once in the car, adrenaline took over. My foot went to the floor. I spun out the drive, around the corner, and up the gravel lane trailing a cloud of dust. The vehicle seemed to go airborne as I crested the second hill. There was a clunk, and I coasted forward until I sat right in front of the Millers' place. The engine was dead. I tried restarting it, but nothing. I vaulted from the vehicle to rouse the house.

As Naomi got her things together, the boys prepared a buggy.

By the time we got back to the cottage, I felt as though three lifetimes had passed. I ran into the bedroom and found, with unspeakable relief, a living, breathing Mary. No baby yet. In my absence, wonder of wonders, things seemed to have stabilized. Mary was calmer and the contractions somewhat milder.

Naomi asked me to heat up some water, and when I returned she announced that I should fetch Edna.

But I had no car.

I ran across the street to the "English" neighbors and pressed the bell. In a minute I had the keys to a borrowed Oldsmobile and was careering down the highway, having learned nothing from my previous mishap. At one bend in the road I almost had a head-on. I roused Edna and drove a little slower on the way back.

•   •   •

The next eight hours were, in truth, our expedition's greatest single test of physical and psychological endurance. Edna, Naomi, and I sat on the bed trying to alleviate Mary's pain. I continued with the breathing exercises and squeezed her arms. The midwives rubbed her legs and feet. A third midwife came to relieve the others, and they rotated sleeping. With each contraction Mary roused from a deathlike slumber as if giving her last gasp.

Childbirth is sometimes depicted as a heartwarming phenomenon of nature, a fuzzy family-friendly photo opportunity. For me it was like trying desperately to revive someone, over and over and over, from a death swoon. Mary is curiously amnesiac about this stage of the ordeal. She tried to bring it to mind later but failed. It was like a bad dream that began to fade as soon as she woke up.

By the time Easter services were in full tilt through the county, Mary's contractions had slacked off to five-minute spacings. Dilation was almost complete. The baby's head was molding to the pelvic opening.

Since we now knew the doctor personally and since there had been no progress in several hours, Edna asked us if we wanted to call him for a second opinion. There appeared to be no present danger, but all things considered, a friendly corroboration would be helpful.

It was too late to reach him at home. By now he would probably be in the middle of an Easter service. Naomi went across the street to call, and twenty minutes later an expensive-looking white car pulled in the driveway. Almost as soon as he came, Dr. Brewster was going out the door again. "Patience, patience," he said. He laughed when I told him what the birth manual had said. The contractions didn't fit the normal pattern, he admitted. But there was no normal pattern.

It was seven hours later, just after six p.m. Something wondrous began to happen. Mary sensed the same thing and, breathing with a second wind, steeled her nerves. She took to the final exertion with athletic prowess. With each push I saw more of the crown of the baby's head. In the aura of the dusky light that was streaming through the windows, the sight at first was strange and vaguely botanical, as if a fast-growing polyp were threatening to despoil her already ravaged anatomy. With the peal of his first *"Waaaaaa!"* it took a second or two for the realization to sink in: here was another person, like Mary, myself, or the midwives, entering the rigors of collective effort. I had no prior inkling . . .

Purple though this alien visitor was, he calmed down quickly, and after his cord was cut, after he was washed, wrapped, and weighed, while the midwives busied themselves getting Mary and the room cleaned up (cheerfully scrubbing blood from the sheets), I was needed as a temporary cradle. So I took his tiny form in my arms—I'd say he weighed on the light side of a pumpkin—and began to walk him around the place. As he panned the rustic quarters, did I catch the whiff of critical bemusement? I shuddered as the words came to me: "Just my luck: no electricity." No doubt my imagination. Still, the encounter more than made up for all the agony. I had gone from wild fear to wild, inexpressible joy.

I thought I was going to lose one, and I gained two.

Over the next several days, Mary and I recuperated and endeavored to smooth Hans's transition to the wilderness. As we adjusted to the new routine, what with nursing and diaper-changing and midnight wake-up calls, our nearest Minimite neighbors on both sides, Millers and Joneses, cleaned up the house, mowed the yard, chopped the wood, transplanted spring greens to the garden, baked bread, washed diapers and clothes, fixed the gate they noticed the horse trying to escape through, and in general thought of what we needed before we did. Mary and I appreciated the extra time to spend with Hans. So, evidently, did Hans. But he seemed most alive, alert, and thoughtful, I have to say, when the whole crew came in to pay their respects. It was as if he were trying to relay a message he just couldn't quite think of. Well, I'll say it for him: "Thank you!"

For all its uncertainty, we were grateful to have had the labor at home. My own sister, a nurse, had a baby shortly later and told a tale of horror in the hospital. Compared with technological overkill, what Mary and I underwent seemed tame. Our only moment of danger, in reality, had taken place when I was on the road in a speeding machine. In retrospect, home birth raised the same underlying question of many of our other bodily endeavors here. Was it labor or luxury?

Just don't believe everything you read.

husbandry

The word house-husband is redundant. Of course! This startling thought came to me as I reached for the hand pump. The "hus-" from "husband" is simply the Old English form of the word "house," while "band" means "bound." The man who stays at home to work is returning to a long-forgotten calling preserved in the language like a fossil. There is no linguistic need to add the extra "house."

Let's see, my thought continued as I carried the two five-gallon buckets of water from the pump towards the hand-powered washing machine. If I stop and set these buckets down here, I can stir the kale and cheese sauce before it sticks to the pan. This reminded me: Oh, I still have kale to hoe in the garden. I don't even want to think about the peas. When I come back, I'll take that hoe and go out and catch up. And then I'll have to walk back to the Millers' to borrow the disk for the pumpkin ground. Musn't forget that.

At this point, Mary interrupted my thought.

"Oh, Er-r-ric," came her sweet voice from the bedroom, "can you tighten the band on the sewing machine? It's come loose."

"Now?"

"Well, I'm trying to get this comforter finished to send off to Karen."

I thought, Well, if I do that now, the water for the washing machine will cool off, so the diapers I'm about to do won't get very clean. Egad, what's that?

The commotion I heard turned out to be Mr. Miller pulling in the driveway. His hoary beard quivered as he drawled, "Do ya wanna dig postholes today?" The deal Mr. Miller and I had worked out bartering my services for rent was working out very nicely. But one never knew when one's services would be required.

I paused for a second. *If I help with post holes today, I can do the laundry quickly first, skip the hoeing, and catch up to where I wanted to be rentwise this week. Of course, digging postholes is no vacation.*

"Sure."

"One o'clock?"

"Uh . . . how 'bout one-thirty?"

"Okay." And he trundled away up the hill.

"Eric, I'm still waiting!" Mary called from the bedroom.

"Be there in a second." *Oh, I forgot about that. If I pour some of this water into the kettle for dishes, I can get it heating while I fix the machine, and when lunch is over I can clean up in a jiffy.*

"Eric!!"

"Right there."

"Can you help Hans? I see he just spit up. Do you smell something burning?"

Simone de Beauvoir portrayed the entire history of male-female relations as one long, willing submission on the woman's part to a monopoly of male power. The evidence in her favor seemed to be overwhelming. But before she ever made this pronouncement, G. K. Chesterton, noting the plight of frustrated working-class housewives in London, observed: "The woman does not work because the man tells her to work and she obeys. On the contrary, the woman works because she has told the man to work, and he hasn't obeyed."

With my recent experiences in the country, I could see Chesterton's point. And anyhow I had long harbored doubts that, in modern times, a tacit patriarchy adequately explained the kinds of privation many feminists would attribute to it. Laborsaving machinery, if anything, had equalized many physical differences between the sexes only so that both men and women could compete to fill roles often more narrow and restrictive in their own way than "traditional" ones. From what I'd seen in the wider technological world, I would not describe the ruling order as "patriarchy" at all. A closer term might be "tetrarchy"—rule by the square-edged. Where everything has to conform to the well-defined slots of a mechanical order, there is no place for something well-rounded, something fuzzier and hard to define—thus no value for what Betty Friedan immortalized as the "feminine mystique." Friedan, by the way, in that famous book told how the call to motherhood quashed her hoped-for career as a professional psychologist. Thus, given the state of the world around her, femininity—or more properly, fuzzily indicated, undercompensated women's household duties—diminished any clear sense of self. Friedan's book might never have been necessary had she grown up on a working homestead. The job description of a subsistence farmer is so mushy and all-encompassing, it would dissolve the self-definition of the modern high-achiever in an instant.

It was not until this morning, however, that I realized the implications of Friedan's pronouncement for my own manhood. The birth of our baby had coincided with one of the busiest times of the year, planting season. And Mary and I had not yet worked out any reasonable division of tasks. One sign of this failure was the frequent claim by one of us: "That's not my job." In a crunch, each expected the other to fetch food from the cellar, heat the water, hoe the garden, pick the summer squash—tasks that before seemed to belong to whoever stumbled onto them.

We took a trip to Kmart for baby supplies. The store was across the parking lot from the mall where I sold my sorghum. When we got to the check-out counter, $127.46 lit up. The cart was laden with a small mattress, cloth diapers, baby blankets, and an infant car seat. Mary looked at me. I looked at Mary. Didn't she know that baby items were the mother's responsibility? "Can't you use your credit card?" I asked.

It wasn't helping that neither Mary nor I had had a decent night's sleep since the birth. Hans, unfortunately, was what's known as a "night nurser." We gawked with envy at those mothers in the community who held their sucking infants tight to their breasts during reasonable daylight hours, then laid them gently in their cribs at dusk, contentedly glutted on warm milk.

Hans was alive and alert the whole day long—no time for such details as nutriment or maternal comfort. He was a serious boy, an industrious boy, a boy who would go far, *a boy who would not waste time on trifling matters.* He'd rather reach for the mobile in his crib, cast a glance at the slightest movement in the room, shake his rattle. Only when there was nothing better to do, maybe about nine or ten p.m., would he relent and accept the inevitable—the long, monotonous task of feeding.

Turning to formula was out of the question. For one thing, the expense would have bankrupted us. For another, it violated every principle we came here to test. It was a technological redundancy. Besides being already available and free, breast-feeding was healthier for mother and child. Feeding on demand is also nature's own contraceptive, and we felt strongly about not using chemicals or mechanical gizmos to space our children. (Since Hans, they have come every two and a half years.)

But we had to acknowledge a tradeoff, at least given a boy who nurses mostly at night. The problem was compounded by a certain mistake we made. A woman we know, who was into "natural" methods of child-rearing, told us about "the family bed." "All the literature I've read," she avowed, "says it's the only way to go. In Third World countries, that's how everyone sleeps. It's just natural."

We were now lying in the bed she had made for us. Hans was not only a night nurser, he was a night wriggler. By morning he had cleared for himself a broad field equivalent to about half the surface area of our full-size mattress, leaving Mary and me each a sliver on either side. I didn't really doze off until just about the time the rooster crowed.

Since the "family bed" had become an immovable plank of our natural creed, we didn't really understand what was happening, why I in particular was getting grouchy. (Mary could sleep through anything.) Sleeplessness must be "natural," so the problem had to be my lack of adjustment to being awake twenty-three hours a day. Over the years, as we gained more children, we unquestioningly enlarged the bed to make room for them—until finally one day, the parents' section mysteriously traveled to another room and I was able to sleep again. That was the day I truly awoke from our friend's advice.

—

Despite some of the dislocations, we were getting what we had asked for when we came: physical and psychical metamorphosis. Feeling good by feeding one's own needs was no longer possible or important; the satisfaction came in seeing someone else's met. Out of the short-term chaos of bodily and chronological rearrangement was emerging a new kind of order, a literal extension of ourselves with its own semi-independent center of consciousness. From the ruptured chrysalis of our former habits a new butterfly was emerging.

One day we got in our car and took a twenty-five-minute ride to a red brick building set halfway up a hillside in a copse of pine trees. Mary's parents had driven down from New Hampshire, and a friend of mine had flown in from Boston. And there before the altar of this small Catholic mission chapel, a man in a Roman collar committed Minimite heresy: we had our baby baptized. Then we returned home and celebrated.

The trip had one other odd benefit. We learned that the purring of the vehicle put our baby to sleep. This environmentally incorrect gas-guzzler was the greatest pacifier since mother's milk.

Corrupted by this illicit discovery, we extended the trip. We felt we deserved a little sweetener. We took the Escort for a long weekend. While Mary attended the baptism of a niece, I detoured to Elizabeth-town College near Lancaster, Pennsylvania, to hear talks on the history and social relations of the Amish.

What a shock it was to experience academia again. I thought it might be recreational for a day, but in the second session I sat in disbelief as two scholars, back to back, cleverly applied feminist interpretative canons to the Old Order, liberally sprinkling such terms as "gender" (a socially constructed "sex"), "patriarch," and "male order." The case was plausible on the surface: Amish men enjoyed nominal leadership roles while women may not even speak at church councils (although they may withdraw assent by not participating in communion); furthermore, there is a strict delineation of "male" and "female" tasks on the farmstead; finally, men are considered the heads of the households. The Amish appeared a lingering bastion of codified male domination. Thus in the feminist scholars' eyes, it was no surprise that from time to time Amish women, to varyingly covert degrees, had resisted.

But I was fit to be tied. Neither of these scholars had actually lived with an Old Order people. I raised my hand. "You have made an excellent case to advance your points"—I was actually granting more than I should have here—"but have you given any attention to the opposite thesis?" (Eyebrows arch in consternation.) "Namely, that never was there a society in which female, or womanly, values so dominated? Nurturing the land and the crops, deferring to the wishes of others, not having to get one's own way? And because they live on farms, women make an important economic contribution to the home, well recognized by the community. I have spent a year living with the 'Amish,' and when men work in women's departments, women tell men what to do—"

The entire audience was letting out a sigh of approval. The scholars' expressions seemed suddenly to grow tight and obsequious. They knew they had no way out. After all, many attendees were drawn here by an attraction to traditional ways, and several were in the garb. Quickly, oh so quickly, the contention was granted. But this did not contradict their arguments, the scholars said. I felt a rich sense of catharsis, as if something long building up inside needed to get out. Of course, it was not my job to defend every Old Order practice, and some of them clearly are not transferable. But what I felt the scholars were revealing was that a certain degree of separation and definition was needed in a way of life overwhelmingly indebted to the indefinable nurturance of mother nature. There was a niche for a man as well as a woman.

Mary and I turned back to our farmstead, newly invigorated in the course we had taken. With me at the wheel, Mary looked for landmarks and Hans slept peacefully in the back. We were all *hus*-ward bound.

pulling in the reins

Having divested ourselves of washer, dryer, refrigerator, microwave, electric lights, computers, air conditioning, central heat, power mower, and running water, Mary and I had come down to one major modern item: the automobile. There was nothing else to take away. We were restricted technologically in the most intimate quarters of human intercourse, but we could step in a car and gallivant across country on a whim.

I suppose that at this point I could have continued to make the argument that as outsiders with special concerns and needs, we required a certain mobility which the rest of the community could just as well do without, and that "use in itself does not constitute abuse." But given my purpose in coming here, there was a more important consideration. The time in which we could test Minimite technologies was rapidly slipping away. Going without cars was not merely one example of minimation. It was perhaps the premiere example, the choice on which the others in some way depended. The slow pace and limited travel sphere of the Minimites was integral to their whole way of life. It was the precondition of neighborly stability, mutual aid, and everyday face-to-face interaction. Being car-free also made them more carefree, financially speaking, more solvent.

What were we waiting for?

There was something a little scary about letting go. The car was our last thread to our former lives. No more possibility of impulsive flights, last-minute outings, frenetic bridge games. To switch to horse and buggy as a primary means of transit was to commit to the settled life of the neighborhood; it was tantamount to becoming Minimite. We didn't know if we were ready. To outsiders we would certainly appear to have made the change; to insiders we might open the door to misinterpretation.

Then we learned of Mr. Bernhardt. He was possibly the most free-thinking Minimite in the area. Although raised Amish, he despised organized religion and never formally joined any of the Anabaptist communities with which he had associated, including this one. Mr. Bernhardt was a bit of an experimenter. He had taken his family across the Gulf of Mexico to Belize, where he ate mangoes and dodged scorpions in an Old Order Mennonite enclave for a dozen years. He said he would fly in an airplane if he ever had reason to do so. And one day he decided to get a driver's license.

So he drove his buggy into town, parked by a tree, and went inside the Registry of Motor Vehicles. The lady at the counter gave him a fistful of reading materials and told him to take them home and study them. He did this.

He began to peruse the booklet the clerk said he must learn cover-to-cover to pass the written test. It was about thirty pages of densely packed road sign pictures, statistics, regulation distances, speeds, and times, and a confusing and seemingly overlapping hodgepodge of driving rules.

He finally stood up, squinched his nose in disgust, and solemnly proclaimed to his kibitzing family, *"Too Much Responsibility!"*, and with a flick of the wrist tossed the pamphlet into the woodstove. I could almost picture the worry lines on his face smoothing over to the pleasant crackling sound of that burning booklet.

Maybe our fears were misplaced.

I admit Mary and I may have needed a reminder of what we and other members of our society had taken on behind the wheel. The responsibility has crept up gradually over a nine-decade cultural metamorphosis, almost imperceptible to those engaged in it.

We began with clunky little Model Ts that barely outstripped horse travel, and we ended with a freeway commute in eight lanes of traffic—often still barely outstripping horse travel. I remember picking up a book called *The Western Way of Death*, written by a British researcher who had measured the stress of driving on motorists. The minute you turn the key, your adrenaline levels rise alarmingly, even if you feel relaxed. Since the only way this stress chemical can be used is through exercise—biologically it serves the "fight or flight" response, impossible for someone strapped in a moving vehicle— over time it leads to fat deposits that line and block the coronary arteries. The worst pollutants the car produces are inside the body. Even as it clogs the arteries around us, it blocks those within. The car is a kind of universal coagulant.

Before we got around to an actual decision, though, a sheriff's deputy, all polite and smiles, appeared out of the blue to inform us that he had seen us in our car with the out-of-state license plate. Would we please update the registration to conform with our present place of residence?

Mary had bought the car in New Hampshire and had kept it registered there, where her parents lived. New Hampshire is one of the few states that does not require car insurance. That, I will confess, was the reason Mary registered it there. Massachusetts, her true former home, required insurance and the rates were sky-high.

The state where we presently lived also required insurance. We stopped by an Allstate Insurance office to see what local rates were.

We walked out of the office in shock. It would cost us a thousand dollars for the first year to begin insuring a previously uninsured vehicle. We didn't have that kind of money.

Our transportation dilemma thus resolved of itself. It was like a case of acne that disappeared for lack of money to buy candy.

The sale of the Escort brought $2,150. With this, we approached the local blacksmith and horse trainer, a soft-spoken man with massive forearms like the smithies of yore. My discussion with Wilbur had confirmed doubts about a massive horse investment. But in a community built on a horse-and-buggy scale, a horse was the standard means of transportation. The ideal choice for us would be a single general-purpose work animal. The blacksmith had one for sale. She was a young mare, who, he assured us, was gentle and did not spook in traffic. At three years old, she was on the small side and easily managed. She could pull a cultivator in the garden or a small implement in the field. He brought her out from the stable, and she stood before us holding her front legs together coquettishly. The blacksmith's son, a wide-eyed twelve-year-old blond, peered through the dangling leather straps as I dickered with his dad on the price. The horse didn't have a name, so after leading her off the premises, we gave her one that suited her feminine demeanor: Isabel.

Next we visited Elbert's buggy shop. He showed us a used spring wagon in good repair, with a second, removable seat that could be placed in the cargo section.

After paying $700 for Isabel, $650 for the wagon, $150 for the driving harness, $70 for the work harness, and $200 for a couple of used saddles, we were still almost $400 ahead from the sale of the car. Had we done something dishonest? We had made out like bandits. And the engine of our new cart, being female, was self-replacing.

Like magic, horse and buggy were in our barn. Could they really belong to us?

I had driven a buggy once in Lancaster County, but that was a long time ago. Now, as I picked up the reins, the first thing I noticed was the slight delay of the horse's response. Because of the rapid speed of a spring wagon compared to a plow, I was prone to over-correct. On a wagon seat you sit very high and feel tippy—as if, at a bump, you'd rocket out into space. At the first slope, I nearly panicked and slammed down on the brake as we accelerated to the tremendous speed of fifteen miles per hour.

The wind gusted, and I felt a sprinkle on my face. I blinked, batting the droplets from my eyes. But there was hardly a cloud overhead. Where was the moisture coming from? Then I realized. From the horse. There was *no windshield*. I couldn't help laughing.

And I relaxed.

It became a joy to reenter the world of sensory objects that is cut off from the driver of a car. You could actually *touch* the hedgerows slowly moving by (the driver sat on the right). You could literally *converse* with the human figures in their yards. The wagon was a kind of moving front porch, and we became transitory neighbors to every household we passed.

To be sure, we knew that some of the onlookers—not all were Minimites—probably said to themselves, "Ah, so they finally did go off the deep end, just as it looked like they were going to. Tsk, tsk." To those skeptics, we wanted to shout back, "We aren't doing this on a romantic lark! We are doing this because it makes sense! It is the next logical step in our search for technological mastery, for true human convenience!" We, in short, were seeking the same thing they thought *they* were. But they tried to do it in cars. Most of them supplemented their rural livelihoods by driving to nearby factories and by shopping in the centralized big-box stores that depended on generous motorists like them—who would never stop working, if only to pay their transportation costs.

The horse's fanny waddled in front of us and the world slowly passed. We meandered from the planned route and tested the vehicle's road-steadiness. The driving took less and less concentration. Isabel, at the same time, got to know the ways better. The responsibility became ever more equine. An occasional tug, a foot on the brakes, and she did the rest.

———

Owning a horse and wagon closed the distance for us among Minimite farms. Without a car, life miniaturized. We began to experience the community as a sort of village and to grasp its real structure from a villager's standpoint.

The dental apprentice, Nate, was one farm behind us. The saw sharpener, Cornelius, lived three farms down from us. Next to him was the retired bishop who sold old-fashioned galoshes. Catty-corner was a farmer named Isaac, who marketed vegetable seedlings canning mason jars and sorghum molasses. Beyond him were Elbert's buggy shop, Edward's furniture shop, Wilbur's harness shop, Jim's general store, and Arthur's feed mill—all contiguous and connected by short-cuts and backwoods paths. What appeared to be a backward farming district was a complete retail complex, with its natural equivalents of skywalks and escalators. No one had planned it; the whole had organized of itself, much as cells in the body differentiate to form a larger organism.

If the convenience was everywhere palpable, it was also surreptitious. New relationships and connections were taking place all the time, but in ways that could not be controlled or foreseen.

I happened to meet Howie again one day at a house raising. As he shadowed me in the attic and pounded in floorboards, bits of friendly conversation filtered through the din. Reminded that he had organized a pepper consortium among fellow Minimites, I asked if I could participate. On impulse he said yes—then bit his lip. It was clear he wanted to take his words back. For an outsider to enter into economic partnership with confirmed members of this group was, shall we say, irregular. But the integrating forces of mutual contiguity literally outstepped forethought.

The result: during the second growing season I expected to add peppers to my cash crops.

The occurrence was not atypical. As if participation in mutual aid were not enough—here a house raising, there a harvest bee—each get-together seemed to provide the hinge to another economic opportunity. Connecting the dots by cloppity-clop. As the pieces came together, our livelihood almost seemed to be generating itself.

I was clearheaded enough to realize that, as a form of travel, the horse was not universally appropriate. For farmers, or others living in rural areas, the horse made sense. Besides shrinking the distance among farms, a work animal provided manure for the fields and power for the implements. But city dwellers might view their situation differently. For them, the human leg makes sense as a primary mode of propulsion.

We, however, weren't sure we'd be returning to the city.

News reached us of a farm that had come up for sale in the area.

One of the Minimites was relocating to another part of the county, and he needed to unload his place. It was outfitted for Minimite use and conveniently situated on one edge of the community, not too far from the county seat.

On one of our excursions, we stopped by for a look-see. It had a working well with good drinking water. It had a large cistern that collected rainwater from the metal roof and fed it directly to a pitcher pump at the kitchen sink. The house was ample in size, solidly built, and well laid out. Several outbuildings surrounded it: a barn, chicken coop, equipment shelter, and large workshop.

The view was nice. The property was not too big, not too small. It had enough pasture for a horse and a cow, enough tillable land for a few acres of pumpkins and sorghum, enough woods to cull for firewood. There was even a small guest house in case we had visitors or wanted to attract future neighbors. Nothing was electrified, and the toilets were outdoors. The price was affordable.

After the success of our crops, we were in a solid financial position. Our costs had gone down while our income grew. Mary had enough savings from her old accounting job for a down payment, and the price was low enough that we didn't think we would have any trouble reselling it if we needed to. It would be a simple matter of signing a few papers and handing over a check. Even if we weren't sure we wanted to stay, we could hold the property temporarily until things became clearer. Properties like this didn't come up very often; it could serve as a hedge against future uncertainty.

After looking at a few more farms, we realized the place was a steal. We signed the papers. It provided a hedge.

———

As our time to make a decision approached, so did the community's. The population was bursting at the seams, a second schoolhouse was being built, and fresh leadership was needed. Much was at stake. Candidates' personalities and inclinations were by now pretty well understood, and everyone silently considered the choices they represented. I wasn't sure what the choices meant to them, but to me they might well frame the identity and character of their whole venture. What, then, were they about? Mastery, in the subtly yielding sense in which I understood it? Or servitude, blind conformity, mechanoid motion? True wholeness, leisure, freedom? Or compulsion, extremism, penance?

The day at last came. Members assembled in the meetinghouse. Prayers were invoked; words chanted. Ancient tradition had determined the method. By biblical precedent, nominees were officially designated and straws were presented. As once long ago Matthias and Barsabas had stepped forward to fill the place of fallen apostle Judas Iscariot, so this day did Wilbur and Edward make their way to the fore of the assembly. From Luther's German Bible, each drew his straw. The results were plain for all to see. The choice was made. It was Wilbur.

And so, in the race to become the next Minimite minister, the son of a converted agribusinessman from North Carolina vanquished the progeny of a Cleveland corporate executive. A community of backward-seeming yeomen that entrusts its future to serendipity cannot be all bad.

Then again, I suppose one cannot rule out the complicity of some outside agent, that hidden presence to which the Minimites in fact gave the credit.

Mary and I were out-of-doors one day when we saw a bicyclist approach from the crest of the hill. I recognized the old-style suspenders and straw hat and wondered what a Minimite youth—and he seemed young—would be doing on a bicycle when to the best of my knowledge, the device was not in favor here. As the cyclist neared, I recognized him and my jaw dropped.

"Bill!" He stopped and we stared at each other. He looked unusually glum. Finally I asked, "What are you doing?"

"Just came from town."

"Town? What were you doing there?"

"Working at my new job."

This was very strange. Minimite members, even newly weaned urbanites like Bill, normally did not take city jobs away from the farm. This could mean only one thing. Bill was no longer—

. "Tell you more about it some other time." He pushed off and disappeared down a side road with a little plume of dust trailing behind.

"Who was that?" Mary asked. I told her the whole story, or at least the parts she had not already heard.

Bill's father, I knew, had returned to the area. I found him on his front porch. He was easy to recognize; he had Bill's willowy body and darting eyes. I identified myself.

"Ah, yes." He smiled. "Bill's told me a lot about you."

I mentioned why I had come.

Mr. Richards looked downcast. "It's been a hard year for Bill. Nothing seems to have gone his way."

I found Edward in his furniture shop and told what I knew.

He turned to me with a look of wrath. "You and Bill both. It's a real effort. I didn't really want you here." The outburst left me reeling. Then Edward sank back, sighed, and closed his eyes. "It's not"—he smiled and nodded his head—"that I don't want you. I knew I had a hard nature to get along with—"

In confusion I quickly interjected, "You have a nice nature."

"There you go telling a fib."

"You have a nice nature. It's hard some of the time." I was becoming flustered and was clutching for words.

"Wh—"

"Being in the army all that time. They really indoctrinate you"— my voice had begun to shake and I was running out of things to say—"brainwash you into giving commands. It's ingrained in you."

"Ingrained in me by my father."

"Oh . . . That could be."

Edward divined my agreement, and reared back again. "We've been through this over and over. Bill has to learn to submit to authority. That's just the way this community works."

Did Edward really think the community's will was concentrated and embodied in him? Did Bill have the same problem of "submission" with other people here? "Look," I said in indignation, "all I know is he's a very pleasant guy to get along with. I know people who, yes, I'd say are not cut out for this. If you brought them here, I would say what's the use of trying. But Bill's not like that. I can really sense how he feels. We have a common . . ."

"United against a common enemy," Edward replied drolly. He was in retreat again.

"No," I lied. I groped for a more diplomatic expression. "It's just the same feeling, maybe, between employees in a large company united at the same level of management. It's just natural for them to talk about their experiences." This analogy didn't seem to help matters.

Edward's voice swelled with pain. "I've seen it happen before like this over and over again—outsiders who think they can come in. It's always a miserable flop. He has to learn to submit to the authority of other people. It's always a miserable flop."

Always a flop? Was he forgetting the many outsiders who had successfully entered the inner circles of the society? Or was he speaking for himself? Had his own transition been a flop?

As all of Edward's pretenses toppled before my eyes, he suddenly seemed small and pitiable.

Grace was in the room now, and he went on, "I guess I'm disappointed mainly in myself." His head was turned aside, and the lamplight caught a watery sheen in his eye. "I know I can't say someone else caused or manipulated me into doing it. That I have to place the blame on myself. I'm just"—his voice cracked—"upset with myself."

A long stiff silence followed.

"I've been a bad example sometimes too," I said meekly.

"No, no. I'm just upset with myself. I guess I should be thankful to have light shone on my weaknesses."

Grace soothed him, "Now, now, we should all be thankful for as much."

I began to blurt nervous filler words before I could stop myself. "Anything could have caused it. It might have been just a convergence of circumstances. I don't know how many times I made a complete social idiot of myself, in front of lots of people, when the reasons for what I did seemed so unfair—five papers due that week. I was twisted out of shape in Boston—"

"Soooo. Now I'm a social idiot."

Grace quickly soothed him again. "No, he's just trying to say he felt like he was a social idiot when he was under pressure."

"I wasn't born yesterday."

"No, he was just trying to say what he felt like, so you would feel better."

But I could already see Edward had begun to take my *faux pas* in fun. There was a mischievous glint in his eyes. Warmth entered the room. We luxuriated in a long silence. Each of us rested in that warmth. No one wanted it to leave. No one dared move. There was at last, for busy Edward, an unscheduled pause. But there was a certain sense of incompletion too. None of us could think of the right comment or acknowledgment. The timing was off or something. I realized nature was calling me, so I excused myself and went to the outhouse.

The air outside was brisk and the sky above infinitely deep, black velvet swathed with celestial dust in layers folding forever back on themselves. It looked like a cosmic handkerchief after a divine sneeze. I halted in my steps. Weirdly, I began to feel myself drawn headfirst into the folds of the galaxy, passing through the heavenly orbs and merging with the velvet blackness. I was on my knees now, and my head was spinning as I dissolved into this beautiful cataclysm. For several minutes I swirled in ecstasy.

Something had come back to me, something that Grace had confided about their past. It hadn't registered when she told me, or I hadn't really cared to hear it because it didn't fit in neatly with my theory, the stark psychological profile I had been forming of Edward. Before moving here the Pendletons had sold their farm in another state to neighbors who didn't have enough income to qualify for a bank loan. So an agreement was reached for good-faith payments—which never came in. The Pendletons had no legal recourse and wouldn't have pursued it if they had. They didn't believe in suing.

Their whole farm, everything they'd worked for, was gone.

What occurred to me now was that this explained much about Edward's behavior. He and his wife had had to start all over again when they moved here. And even with that disadvantage, they had taken on the responsibility of adopting two small children with big medical problems. (One time in an ice storm four-year-old Amanda had gone into a grandmal seizure and had to be conveyed to a hospital twenty-five miles away. The bill was five thousand dollars.)

By leaving themselves so vulnerable, it appeared, yet again, that the Pendletons brought on their own suffering. But in doing so for the sake of others, however rashly, Edward and Grace appeared united in the choice that until now I had overlooked: Christian self-sacrifice. Of the two of us, Edward and me, it was not clear who had been harder on whom. If I could read Edward's hardness in his actions, would he not have read, and suffered, my hardness in mine?

When I returned to the room, they were still sitting by the light. Slowly and sincerely I spoke: "Just one more thing while I'm here. You had mentioned something about feeling you'd been a bad example. I wanted to say that if there was any guilt about that, I forgive you."

"Oh, I—"

"And I know that I may have been irritating and absentminded and probably done other things wrong. Will you forgive me?"

"Of course. It was me I was mainly upset with." As Edward talked and protested, then, as though from hidden recesses in the room, something like honey began to flow thickly, filling the cavities with sweetness. It enfolded us all in a soft embrace. The room was so still, we heard the sound of our own breathing.

Now Grace was talking: "You know how a person can coast along and think 'Stress never bothers me,' and then it comes along and you go like—" She made a face like a grizzly bear's and held her hands like claws.

I nodded, adding, "And you had to work with Bill and me at the same time. That must have been like having to work with Laurel and Hardy."

"I think"—Grace was at her best now—"you've both demonstrated good Christian behavior tonight."

Edward now was lying across the couch in an ungainly sprawl, hat off, his hair flaring in all directions like the corona of the sun, smirking and feeling deep in his pocket. It looked like there was a live squirrel in there. "I hate it," he said, "when there's a hole in my pocket. It makes be feel so *ulglglgh*. I can't seem to find it."

Whatever had slipped out, I hoped it wasn't anything he'd miss.

We saw Bill one last time. He was in one of his punk-rock getups, speeding by in a beat-up Oldsmobile, his hair long and flying.

TWENTY-ONE

closing time

October winds ushered in a swirl of change, and everywhere brilliant orange and yellow leaves descended to the earth. Our second batch of crops came in more plentifully than the first, and the profits included the proceeds of Howie's peppers. I had prepared the ground this year with my own horse and borrowed equipment from the Millers.

And let me not forget our other little project. With constant care and a rich liquid diet, he now weighed in on the heavier side of a pumpkin, this labor of mutual love, Hans.

With autumn, the time of our own decision had come. We were now three, and the month we had entered was the eighteenth. The results of our tests were in, and they were favorable. All except one.

Mary had been accompanying me ever less frequently on the buggy trips. Something about them seemed to bother her. It started as watery eyes and a runny nose. From there it developed into full-scale sneezing episodes whenever she got on the wagon seat. Was she allergic to something? The last time she remembered symptoms anything like these was as a child, when the wind blew into her bedroom window from the general direction of a neighboring farm. It was a horse farm.

Mary had always wondered whether she was allergic to horses. Now there was no doubt.

It was hard to believe that something so small could have effects so big. Horse dander. Mary could not abide the emissions of the prevailing mode of transportation here. We could see no easy way around the problem. Mary was allergic to her best hope for mobility in the community.

We sat in the living room weighing the revelation, measuring it against other considerations, striving for dispassionate analysis. The scales seemed to tip back and forth ever so slightly, but always, inexorably, tilting to the same conclusion. If horse dander had been the only issue, perhaps we might have tried to make some creative adjustment. But we had long been hovering in the balance, and other considerations also added their weight. There were still certain personal requirements we could not meet here. The irony was that they turned out to be mostly mine, not Mary's. My original reason for coming was to prove a point, not to stake a claim. I was restless with the desire for further exploration, hoping to translate what we had learned here to the outside world.

As we were sitting in the living room, from Mary all at once came a burst of sniffles. I looked over and saw that her head was buried in a handkerchief. Allergies? I didn't think so.

I put my arm around her, assured her we'd be okay, and tried to think of the bright side. I pointed out that now at last she would have more time to be with her mother.

"Well . . . But I hardly ever had time to get together with her when I was in Boston even though she lived only an hour away."

"True," I added unthinkingly. "She spent almost a whole week here with you, both times."

"Yeah. . . And that was more time than I've spent alone with my mother in my life."

Another bout of tears was unleashed. When she was done crying, Mary looked at me open-mouthed, like an accountant who has just discovered an entirely new principle of bookkeeping. She spoke quietly: "We *do* have more time here."

Mary's head fell back on my arm, and I pondered the last piece of evidence. My quest to discover how little technology was actually needed for human comfort and leisure was now over, and I believe I had an answer, so far as could be told here: no more than the

Minimites used. Maybe less. Indeed, had we decided to stay, we would have made some adjustments in our own household, tending to less technology overall. We would have sought some form of non-motorized running water. We would have broadened our focus in gardening from summer crops to a year-round rotation, spreading the harvest, à la Eliot Coleman, across the seasons and reducing the need for canning. We would have explored more reliable means of cooling food, possibly using a solar cell (one technology the Minimites did not employ.) We would have opened the kitchen to south-facing light for the sake of winter cheer and solar warmth. On reflection, much worry and danger during childbearing would have been saved with better access to a telephone—whether our own or by bartering for use of a neighbor's, we didn't know. (It would have been nice had the midwife also had better access to one!) And lastly—and least "technologically"—we would have tried to get by without horses, or kept our usage to one, thereby drastically lessening the need for land, labor, and machinery—and handkerchiefs.

Yet even without these changes, the life was good enough. As Mary's mother is our witness, we had more time. This was another way to say that we had fallen *in* time, taken *our* time. The relaxed rhythms of manual labor, like some unseen conductor's beat, coaxed into synchrony the oddest array of harmonizing parts. We had drummed our wooden spoons against the kitchen kettles, mingled with the brass and winds of the barn animals, and soared cerebrally to the accompaniment of string beans. And after much arduous polishing and practicing, we had finally struck a chord with the whole collection.

The secret lay, much as anything, in simultaneity. Things that technology had separated were reunited. The results were more than efficient; they were symphonic. In an orchestral performance an oboe warbles beside a viola and the two produce a lush blend. On the porch of a working household, you visit with your mother-in-law while pushing the centers of tomatoes into a bowl, and the breeze brushes against your face, and the leaves rustle—and likewise music emanates. And when your part is done, there is plenty of time to breathe during the rests.

The marvel of such unplanned synchrony is one surely worthy of a student of economics, perhaps even of an addendum to Adam Smith's treatise on the division of labor. But we had learned the lesson just in time to leave; and we now knew it went far beyond efficiency. We had found a home here. The attachments defied any rational accounting: our cozy cottage nestled under the hill; its airy interior and porches; the way the sun set behind the trees; the way the neighbors dropped in constantly and unexpectedly; and the neighbors . . . our friends . . .

I returned Isabel to the blacksmith. We arranged to have an acquaintance from town who owned a pickup truck haul us and our belongings to the nearest car rental agency.

Then the time came for goodbyes. I could feel a kind of fuzziness overcoming me, a dread at what was to be. I was not good at this sort of thing and tended to go numb. When we saw the Joneses for the last time, Carol and Mary wouldn't stop talking. Nate wouldn't either, even though we had recently had a small falling-out over scripture. But now the stream of verbiage was oddly welcome; I couldn't get enough of it; it filled the gap. Like Edward, but in a more benign way, he was a Bible-worshipper. But he was also a person with a big heart, and after so many hours hauling sorghum and hammering nails with him, I realized more than ever how little it mattered that our opinions differed. Mary and Carol shared a womanly hug, and Nate and I shook hands. Before I knew it, the farewell I had dreaded was over, and I wished I could replay it.

Our last moment with the Millers came suddenly and shockingly. They stopped by to collect the keys, looking polite and crinkly-eyed and a little hesitant. We stood facing each other. No one knew what to say. Here were our beloved mentors who over so many months had adopted us into their circle of mutual aid, guided us through the perilous passageways of agronomic success, and attended to our every smallest material insufficiency. Impulsively, Mary ran up to Mrs. Miller and gave her a hug. Though she reciprocated, her expression hardly changed, and her posture remained stiff. Then briefly I saw her eyes soften and tear up.

Mr. Miller and I shook hands.

"Can't thank you enough," I said.

"Don't mention it. You'll do well wherever you go." He looked at me steadily as he spoke, and I knew that he meant it. It was a kind of fatherly blessing, and the effect was long in rubbing off. The words filtered into the center of my daze and melted it away. He was right. Life offers more than one option for doing the right thing. I let go of Mr. Miller's hand and stood back.

When it was all over, we climbed into the awaiting pickup and, with a turn of the switch, pulled away from our fond habitation, propelled by a sneeze. But a brief gulf separated us from modernity. A curtain of rain descended, like the one we had driven through on arrival. As the little world receded into obscurity, it was hard to imagine we had spent over a year in its green and compact byways; harder still to picture the power this little world might flex upon the larger one which we now re-entered when, given the right finesse and a little luck, the results of this experiment were set at large.

outside, and the box

Back in Boston, I hastily assembled my findings in a rather stodgy master's thesis. It was accepted, though with reservations, and then Mary and I went to work. The exploration I described in the thesis had another segment that was still under way and could be completed only in another setting that it was Mary's turn to choose. In that place, we would try to learn to adapt principles of minimation to new circumstances. We now had a strong hunch where we would end up, but we wanted to save for the move. We had taken to heart the Minimite example of living debt-free. We hoped to make an outright purchase.

Mary went back to accounting part-time, and I, wooed by a friend who owned two taxicabs, got a hackney license. This was a true test of my flexibility: in a city saturated with musicians, I realized cab driving paid better than piano playing. It was only to be temporary, plus I biked several miles each way to pick up my cab. But I soon discovered I loved cabbing in Boston. What better way to get to know the innumerable twisty streets, the colorful historic neighborhoods, and that ever-elusive Bostonian? I could pick up a slice of the best Italian pizza in Allston, chat with a Harvard professor on his way to dinner (one night it was Henry Louis Gates going to a Spanish restaurant on the Cambridge-Somerville line), then stop at a pub in Kendall Square and catch up with Mary on the phone while sipping a good beer (technically a no-no), then off to the airport or to the North End, or downtown to take a lawyer home to the suburbs.

While I was having this fun-disguised-as-moneymaking, we rented a cheap apartment in Somerville, then switched to a better situation in Hyde Park, where an elderly gentleman needed in-home care. We lived rent- and utility-free on one side of the duplex he owned in exchange for doing his laundry and cooking his meals. Our income increased and our expenses went down.

So we stole some time for one daring side-adventure: we left Boston for a few months and got involved in forming a rural neighborhood association, patterned after that of the Minimites, with others who wished to regulate the use of technology. The project proved to be a brief but instructive last use of the land we had bought. We learned that in its early stages, a fledgling community is very fragile and requires the right blend of many ingredients, including personal outlooks. It presumes a certain level of psychological stability. If any ingredient is missing, the whole thing can quickly cave in—and thus, like a failed soufflé, did this one. We sold the property, mourning the loss of our last tie to our original experiment.

After the flop, I learned of another kind of movement that has arisen in response to a yearning, similar to mine, for a socially inviting human habitat. It is called "the new urbanism," and its leader, a Cuban-born architect named Andres Duany, lives in Florida. Duany has designed numerous neotraditional villages like Seaside on the Gulf Coast, and Kentlands outside Washington, D.C., patterned on nineteenth century American small towns. Small lots are platted, wide front porches mandated, stores and residences mixed together, and cookie-cutter developers outlawed. The idea is to encourage face-to-face relationships among neighbors in a varied and beautiful cityscape where all the conveniences of life are within walking distance. What Duany has achieved is a consummate balance of nature and artifice: a kind of planning matrix in which sundry and vital unplanned interactions can occur. Living in one of Duany's artfully planned habitats sounded immeasurably easier than starting a community from scratch.

But then I looked at the price tag. These fancy clapboard environments clearly presupposed an income that could only be gotten in one of the high-paying career specialties that had helped bring the demise of the nineteenth-century American town. The developments addressed intimacy of place, but not so much integration of life and work—the frugal practices of self-sustenance, bartering, and manual labor such as we had found among the Minimites. It was silly for us even to think of moving to a Duanian village.

Our favored destination only seemed to look better with time. Housing there, for one thing, cost about one-eighth of that in Seaside.

But I still had to be able to make a living, no matter where I ended up. And Mary would have preferred to have relatives closer by.

—

One muggy evening I was weaving a fare through a typical late-day traffic snarl, and I quipped, "You know, I'd really rather be a rickshaw driver. That way I could at least work off some of this stress." (Stress being a drawback of cabbing—one reason why I could imagine it only as a temporary vocation.) Biking home from my cab was an invigorating but rather inefficient compensation for the long hours of confinement in a motor vehicle.

The fare responded, "You can be. I've seen them in Charleston, South Carolina."

"Real rickshaws, like they used in China?"

"These ones are bicycle-powered. More efficient that way."

I cocked my head and thought a minute.

The final piece of the puzzle fell in place when we learned that a pair of Mary's siblings had relocated to parts of the Midwest not far from the one that we were thinking of.

After about two years, we had saved up enough money for the move.

—

I pulled up in a rented Ryder truck, and not long after, Chuck arrived in his Honda. Chuck was (and is) an old friend from Boston who taught at the nearby state university. Up went the sliding door of the moving van, and we got busy.

Hours later, after all Mary's furniture was in the house, we sat on the lawn sipping beer and enjoying the view. Down below was the village, with its spires and ornate red brick buildings. In the distance was the cupola of the winery, poking through the trees on the facing hillside. "I can't believe you got this for fifty thousand," said Chuck.

"Just what houses go for here," I told him.

Later I got a call from the machine shop. My delivery had arrived, they said. I had it taken there because the crate needed to be set on a loading dock. It was about four feet wide and eight feet long. I headed over and gazed at the enormous container, anxious to view the contents.

"So, what's in the box?" the machinist asked.

I took a crowbar and pried off the front of the crate. Deep in the shadows could be seen a little red carriage with a black retractable canopy and bicycle pedals. Even in the dark it gleamed.

———

"Can I please, please, please?"

I look up from the yard and pause. It is one of the neighbor boys again. He is standing in the alley. He looks to be about eleven. He is chubby.

"Just a *little* while," he pleads.

I sigh. "I need to let you know, I'm not paying you."

"That's okay."

"If you really want, then, you can finish this back section up to the tree."

The chubby boy skips over and I give him the handles of the mower.

Tom Sawyer never had it so good. The motorless rotating cylinder, with its quiet *snick-snick*, has a hypnotic effect on passersby, especially if they are male and under the age of thirteen. I don't have to do any talking—it spins its own spiel. My own two boys, now nine and four, sometimes fight over the handles. They've gotten feisty and require lots of projects around the house to occupy them.

But it's not just the younger ones who become fascinated by this twirligig. Sometimes, in the corner of my eye, I catch a car creeping slowly along, trying to figure out what it is I am up to. Can he really be using a push mower like Pa did?

When this happens I sometimes rub it in, indulge a little, give the product a plug. I turn and smile and say, "You know, this thing's actually easier to push than a power mower. They're made out of aluminum now. And they're quiet as can be. I never have to run to the Quik Mart to gas up. Never had to sharpen the blades in five years; you just adjust these little screws. Only cost seventy dollars . . ."

The car edges away, and I realize I have gone too far. Better to let the *snick-snick* speak for itself. Especially around here. The locals don't like to be told what's good for them. They love all things motorized and metallic. They're different from the Minimites, even though they share a common German ancestry. As the tee-shirt logo truthfully says (for sale at the local gift shop catering to ogling tourists), "You can tell a German, but you can't tell him much." No German-American from this town has ever taken a ride on my rickshaw—only newcomers.

It's the other newcomers here who are more likely to try something "old." A family we know, who moved in from Michigan with ten children playing violin, cello, piano, and sundry other instruments, now lives on a ten-acre farm six miles outside of town. When they discovered we made soap, they began bartering some of their surplus cow's milk for it. I refrained from telling them about thirsty pigs. Another couple, now raising goats and chickens in preparation for the husband's imminent airline layoff, swaps us eggs for soap. When they come by, we invariably get entangled in a long conversation on just about every subject—except eggs and soap.

The Germans hate it. I think they are in a 1950s group-think time warp. The woman across the street told Mary that her mother, of local descent, had a quilting frame she could borrow. The day came when Mary needed it. The woman told her mother, and her mother refused to lend it out.

"That man doesn't have a job," she said angrily, or words to that effect.

I don't. I don't pull a paycheck from a gas station, a factory, a winery, or a teaching faculty. I am alone here in joblessness until the day my airline pilot/goat-breeder friend loses his. I am living more like the people did a generation or two ago in that cranky, amnesiac grandmother's childhood. We simply don't need much money, or miss it.

All this bothers many of my neighbors (maybe they've caught up to 1970 now). But there are enough newcomers here who are at the next stage. They know what the world out there is like; they've had enough of high technology and the corporate job market. We find plenty of company.

There's Pieter. A local organic farmer who's trying to wean himself from his job as a state agronomic researcher, Pieter is adept at using the system to break down the system. He landed a grant to begin a business selling wildflowers from his fifty-acre roadside plot. The road is the main route from the interstate highway to this town, though relatively quiet and untraveled. Pieter let me borrow his rototiller in the spring to plow under a strip of crabgrass along the alley where I planned to plant squash. It was too impractical to use a horse and the roots were too thick and interwoven for a hoe. So I succumb to twenty-five minutes of motorized gardening.

In return we have had Pieter, a bachelor, over for dinner and drinks many a time. Pieter is a vivacious conversationalist, exuberantly opinionated, and generally a great deal of fun. He also plays chess, which I quite enjoy, and we have whiled away many a winter's eve by the fire at this sport.

Pieter, too, has taught me that motorized equipment, selectively applied, is not incompatible with mutual aid. I should know: I get to his homestead six miles out of town in the '86 Blazer my dad passed on to me. I'd bike, but the highway is too narrow and curvy. I ended up accepting the vehicle only because its maintenance costs were next to nothing; the liability insurance in this area, for instance, is only about two hundred dollars a year. We try to use the car only when we cannot walk, bike, or take the train.

Pieter and I are in a card-playing group (poker, not bridge) with Steve. Steve and I also have a relationship that involves selective mechanical application. He runs the copy store. He is also a newcomer to town, and his store is the center of newcomer gossip. To make a single photocopy generally takes me, given the news I must catch up on, at least thirty minutes, and that's if there's no one else in line. It is a very inefficient way to make copies, but a very efficient means to converse. Mary wonders where I disappear to.

But Mary shouldn't talk. When she goes out running errands, there's no telling how long gab will grab her. At first I would think, "She must have gotten hit by a car; it's eight p.m., suppertime is long past, and there's no word." Now I think, "She's probably at the Borzillos." We belong to a food co-op with them (transplants also), and Sally and Mary think and talk alike. Sally loves our soap, but she usually just pays for it. Mark, her husband, is a freelance engineer who cuts his own firewood, built most of his house, and tends a half-acre garden.

I no longer think much about Mary's absences. We spend most of our time at home together anyway. But we have now established well-defined spheres of activity; otherwise we'd trip over each other. The system works well as long as you have a private bedroom in which to rendezvous from time to time. I disappeared for about a year into the walk-out basement, which I transformed into two guest rooms with private baths. The idea was to make our lower level an economy-priced bed-and-breakfast. My Minimite experience hammering nails came in handy. I fine-honed my carpentry skills one summer with a weeklong course at the Shelter Institute in Bath, Maine. I even acquired knowledge of electrical wiring. This town is a crossroads for bicyclists who take a trail along the river, and our B&B has been filled with grateful, paying bicyclists and other travelers ever since. Some of them have become dear friends. We call our place a "Bunk and Bagel." It has turned out to be a greater moneymaker than the rickshaw, which is popular mostly on Saturdays.

While I clunk around with my odd projects, in the B&B, and on the rickshaw, Mary cooks, cans, and gardens. She also gives the kids their lessons. We've become home-schoolers, and with no television, our children have become book-inhalers. But their principal lesson comes from the living examples we provide. At eight, without my instigation, Hans made a wooden mailbox with a hinged door from scrap lumber left over from my basement project, using my hand drill and handsaw. After I traded in my two-passenger rickshaw for a larger one with room for two drivers in front, he became my driver-apprentice. He sits beside me, helping pedal. He earns fifty cents a ride and grins all the way to the piggy bank. Our daughter, Anna, now seven, knits, helps label soap, and can bake a respectable batch of cookies. At three, Evan was already helping push the mower; now, just turning five, he can do half the lawn.

There is a curious way that, after a while, a life with fewer automated helpers becomes lighter than one with more. The dynamics of mutual activity take on their own life and liberate a sense of common cause. There is also a real savings in maintenance on fuel-consuming mouths. Now that we have gotten our routine down, it is a good guess that Mary and I spend only about two or three hours a day on work necessary to our livelihood.

One thing, though, has me scared. The majority of people in this town—only recently emerging from the deep past—want a Wal-Mart. They were delighted when, recently, a Hardee's arrived. To them, the historic charm and convenience is passé. They want a new, wide bridge that cars can take at fifty miles per hour over the river. They drive to the post office, recently relocated to the outskirts. The mayor and city council didn't show up at a single workshop session when the state downtown revitalization organization held a conference here. The city administrator, who appeared for twenty minutes to give opening remarks, used the occasion to make cryptic, apparently demeaning comments. But that wasn't the most distressing thing. The small grocery store in the center of the village has recently decided to become a supermarket on the fringe of town, eliminating the walkable shopping we have enjoyed for six years.

The only hope seems to be a greater influx of outsiders who can see and appreciate local treasures. But they haven't been moving in fast enough.

Fortunately, we never enshrined this place or made it a panacea. What we saw in it can be transported; principles are lightweight and easily carried about. We can bring our religion with us too. The rickshaw may be just the thing to tote them along.

recipe for a leisurely, laborsaving life

The main three ingredients of technological liberation are a pinch of muscle, a sprinkle of wits, and a dash of willingness.

The longer Mary and I have lived and worked in this village, the more we have relied on these three inborn capacities. Pedaling the rickshaw has been largely an effort of brawn; the idea to transform our lower level into bed-and-breakfast guest rooms was an effort of brain. Soapmaking took an initial introduction from friends and has been fine-honed through practice; it involves dexterity at both levels, physical and mental. Only one of my income sources—music—has taken skill for which I trained intensively. I play at occasional weddings and have given a few piano lessons. But this learned ability pays little in comparison with the forms of income we have generated through spontaneous use of body and mind. This spontaneity, this flexibility to changing circumstance, is fundamental to the others, a kind of leaven. It lightens, if you will, the whole effort.

To return to these native capacities is, in itself, to regenerate the human community. Manual effort craves collaboration. As the Minimites put it, "Many hands make work light." Even in a city or sprawling urban area, crafters spontaneously seem to congregate, share ideas, and barter merchandise. I know because I have attended several craft shows in the nearby metropolis and found a thriving sub-culture of creative tool-users. I also found sections of the city, gener-ally older, denser, and more walkable, where persons of this leaning tend to congregate. The areas are gradually becoming more popular.

If Mary and I were to relocate, we might well go there, with little change in our ways. As the future of our small town clouds with doubt, these pockets of low-tech urbanity promise greater vitality: they boast proximity to cultural opportunities our town never had, such as a symphony orchestra. (I have a weakness for music.) Residing in a dense urban pocket is not very conducive to growing one's own food, true, but on our small hillside spot Mary and I have done little of that anyway. We have remained busy enough with our various other household endeavors and can easily exchange the proceeds for fresh food at the store. The neighborhood we admire in the next city surrounds a thriving farmer's market—a visible link to the country-side.

But make no mistake. Whatever the future holds, for the time being our current home retains its appeal. I am constantly surprised by the informal guises of barter. A lawyer I know, whose office is two blocks away, calls me periodically to witness the signing of wills; in return, I can depend on informal legal advice and invitations to lavish fêtes at her seven-acre spread on the edge of town. During a lull in rickshaw business on an Octoberfest weekend, I asked Sherry, the town librarian, if I could borrow a few clips with which to attach a sign to my canopy. She complied—and called payment due immedi-ately with a request to bring her a bag of kettle corn from a refresh-ment stand down the street. Don, my next-door neighbor, sometimes watches over our guest quarters on weekends; I repay him with food and drink, which I deliver in person and help consume on his deck.

Along with the revival of wit must come, of course, wisdom in choosing technology. For those who would outstep and outsmart machines, a broad suggestion: remember the principle of minimation. Technology undoubtedly has, and will always have, some role in mak-ing life easier or better, so one shouldn't exclude it. But the role is sup-plemental. Technology serves us, not we technology.

The principle of minimation can be roughly stated thus: other things equal, it is better to find a non-technological solution than a technological one, or failing that, a less technological solution than a more technological one. There are at least three reasons for this.

First, a modern automatic machine is no mere inert tool. It is a complex fuel-consuming being with needs of its own. It gobbles up energy; it demands care and maintenance; it even has bouts of temperament. In many cases no diaper will contain its mess. And all this on top of the initial chunk of cash it bites—its purchase price—which often amounts to a king's ransom. For these reasons, it not only serves but must be served. But it is more than another mouth to feed; as it becomes more involved and involving, it can easily invade the living space we formerly reserved for ourselves, taking on functions once our own.

This brings us to a second and more critical reason for minimation: avoiding usurpation. To cater to an inanimate object's needs is one thing; to aid and abet our own replacement is another. Duplicating vital human capacities can have one of only two consequences: atrophying the capacities or creating competition between *Homo sapiens* and machine. Neither of these is savory to self-respecting members of the former. This is not to mention the wasted use of resources or the often superior artistry, elegance, or efficiency of human powers or processes.

The third reason for minimation is recalling the need or end in view. A complex mechanical entity readily overwhelms or subverts the very purpose for which it was deployed. This is because of its sheer immensity and the many unforeseen consequences that such a thing may bring about. Blind to this pitfall, the participants in our automobile culture continue to propagate cars, freeways, sprawl, and development that, as Ivan Illich has pointed out, when hidden costs, labors, and time are added in, leave us moving no faster, and possibly less fast, than we were one hundred years ago in horse-drawn vehicles.

Granting these dangers, technology still has an important place. In our experience, it has come in handy in three main areas.

Bodily labor: Over and over again, among the Minimites or here in our new home, I have learned the same lesson. Primitive technologies are often better suited to the task than more advanced ones. In a world of organic beings and relationships, machines can act as a wrench. It often makes no sense to save labor and time when "labor" provides needed exercise and "time" is spent with family or neighbors.

Having said that, I have occasionally found a use for power devices. When I worked at finishing our basement, I mostly wielded a hand drill for drilling or screwing. I liked the opportunity to flex my muscles. I preferred the quiet cranking sound to the scream of the power tool. And I avoided drilling a hole through my finger. But someone gave me an old power drill he was no longer using, so I had a chance to do a comparison. The power drill saved some time, but not that much. For me it was a little faster turning a screw, about twice as fast drilling a hole. Occasionally the power drill was clearly superior. The pressure-treated wood I bought for our deck was very dense, and was hard to drill through manually. Also, in the basement there were some spaces too cramped for rotating hands, so again the power drill was necessary. (A friend recently suggested I try an old-fashioned "Yankee" drill, which has a push-in, pull-out action and can be operated using only one hand. He claims to delight in wielding it, but I have yet to locate one.) My use of a power saw, which I borrowed a time or two from a neighbor, was similarly intermittent. By exercising restraint with these devices, I created a working environment both more serene for myself and more inviting for my young son Hans, who at age five begged to be allowed to drill or saw into his own pieces of wood alongside me. At the same time, the motorized devices could be undeniably handy.

Transportation: By working in and around our home we have saved untold time and cost in transportation. Still, to have minimal need of a car is not the same as having no need. We have gotten by with the Blazer for our longer trips. I have also considered turning to a low-energy car or a hybrid vehicle that can be pedaled but that also has an electrical assist motor for hills or longer distances. Last spring I traded in my three-wheeled rickshaw for a four-wheeled pedal-car called a "Quadracycle," which uses two wheelchair motors and two batteries for its power boost. The front passengers can pedal as much or as little as they like and allow the silent electric motors to pick up the slack.

But most of the time for our short trips in town, it is the simple bicycle that makes the most sense. Mary and I supplement it with attachments known as Trail-A-Bikes, which give our children the pleasure of pedaling behind their parents in temporary tandem. Our bikes have large rear baskets, each big enough for a sack of groceries. I recently replaced my worn-out conventional bicycle with a recumbent, which, with its heavily padded reclining seat, has made cycling more comfortable than sitting in an easy chair in my living room.

Communication: We are selective when it comes to communication devices as much as other technologies. We have a telephone, but not a television or a video player or a computer. The telephone has become so much a part of daily existence for most people that, unlike the true Minimites, we would be socially isolated if we didn't have one. And for my rickshaw, I recently got a cellphone so customers can call me in transit.

In a small town, the need for a phone is lessened. I frequently bump into people I have been meaning to speak to and am able to convey the message face-to-face instead of over the wire. Yet if I want to play a game of chess with Pieter, my friend who lives outside the city limits, I must call.

Although we don't have a television, watching it is sometimes unavoidable, say when visiting friends and relatives. But at these times "tele-voyeurism" becomes an occasional treat, an enjoyable bonbon of entertainment instead of a joyless, lifelong compulsion. When it happens only occasionally and in the presence of others, TV viewing can also become the basis of a mild social experience.

Along the same lines, we look forward to going out to an occasional movie, both for the reason of its rarity in our lives and for the chance to take part in a forum of public entertainment. For certain types of "television content," like news, we turn to newspapers and periodicals. People frequently comment on the fact that our children seem to possess long attention spans—they don't wiggle too much in church and they sit quiet and rapt at concerts and theatrical productions. I always reply, "We don't have a TV at home," and the person who commented, after recovering from the shock, will nod as if no further explanation is necessary.

Mary and I, again, do not own a computer, but that has not prevented us from making selective use of some of its capabilities. We advertise our guest rooms on a bed-and-breakfast website, for example. This ad generates the majority of our business, and does so at less cost than any other form of paid promotion. (When prospective guests want to book a room, they contact us by phone, not by e-mail.) The device that in other applications is an agent of faceless information-gathering, disembodied message delivery, and physical torpor has become a magnet for live interaction with those seeking accommodations in a palpably crafted environment that they can peruse on foot.

Occasionally I do use the computer myself. When time came to upgrade my rickshaw, I took a two-block walk to the library, opened an e-mail account, and placed an ad on the Internet. The buyer was a pedicab driver from New York City, and he traveled over a thousand miles to pick up the vehicle in person. I doubt I could have located this prospect by any other means.

But I resist having a computer in the home. I am presently typing on a device that many would consider stone-age: a word processor. I decided to buy it instead of a computer for all the reasons above. It was much cheaper, it uses much less energy, it doesn't absorb more of my attention than I wish, and it embroils me in fewer other thorny technological dilemmas. And it still does just what I need it to. It is admittedly slower than a computer, especially when printing. But I actually like that limitation. Computers were supposed to reduce the need for paper, but when everyone can print out hundreds of pages in a flash, whole forests topple with corresponding rapidity. Because my printer is slow, I use it only when I am really ready to print, saving reams of paper. And since I have other business to do at home, I can get up and cut bars of soap, or clean a guest room, or play the piano, or talk to my wife, or give my third child, Evan, a reading lesson while the processor prints. It is actually a nice excuse for me to get up and take a break.

There really is no end to the possible uses of technology, nor are there limits to finding a way around it; but in all cases it must serve our needs, not the reverse, and we must determine these needs before considering the needs for technology. The willingness and the wisdom to do so may be the hardest ingredients to come by in this frenetic age. Perhaps what is needed most of all, then, are conditions favorable to them: quiet around us, quiet inside us, quiet born of sustained meditation and introspection. We must set aside time for it, in our churches, in our studies, in our hearts. Only when we have met this last requisite, I suspect, will technology yield its power and become a helpful handservant. Mary and I still turn on the kerosene lamp and read by the fire on a cold winter's eve. By switching off the electric light, I think we see a bit better.

Insights,
Interviews
& More . . .

Amish in St. Louis

Meet
Eric Brende

ERIC BRENDE has degrees
from Yale, Washburn
University, and MIT, and
has received a Citation
of Excellence from the
National Science
Foundation and a graduate
fellowship from the
Mellon Foundation in
the Humanities. At the
insistence of his editor, he
now has an e-mail
account at the local
library but continues to
minimize modern
technology for himself
and his family. The
Brendes have recently
relocated to an old-town
section in St. Louis,
where Eric makes his
living as a rickshaw driver
and soapmaker.

DURING a three-week tour I took promoting
this book, the public hungrily ate up what I
had to offer. *People* and *Parade* covered the
release. *C-Span* aired my book signing in
Cambridge. In between events, as I drove my
rickshaw one night down the street at home,
a police officer yelled from his paddy wagon,
"Hey, I heard you on Art Bell last Saturday
night!" Art hosts a popular nationwide late-
night three-hour radio talk show. The officer
was gone before I could as much as wave.
I was having my fifteen minutes of fame.
And for what?

Everyone seemed fascinated by the thought
of something so radical, so singular, so
countercultural as giving up automated
technology and power for more than one
year. Many people admired my "bravery,"
my "pioneering spirit," my "self-control,"
wondering how I or my wife or my children
could endure such privation. Many called my
ideas thought-provoking and—although
they weren't ready to put them in practice *just
yet* or *so radically*—announced that they
would mull over the possibilities.

But slow *down*. . . . Who is really being
brave, radical, or extreme? Is it I? Or is it
the people who marvel at me? The changes I
have made to live less technologically are easy
compared with the contortions most people
go through to *maintain* technology. Their
beloved machinery does not so much *save*
labor as *separate* it out in time and place and
thereby make it *harder* to obtain—physical
exercise in the gym, moneymaking in the
office, education in the school, and "quality
time" with the family in the national park.

Rather than an integrated whole, life becomes
a temporal and geographical obstacle course.

And even in terms of sheer volume of
technology, I am hardly deprived. Merely by
existing in a Western country, I have ready
access to sanitary water, vaccines, plentiful
food, many mass-manufactured goods, and
select forms of automation, such as electric
fans, a small refrigerator, a dehumidifier in my
basement, and, as mentioned, sometimes a car
and a computer. And with this degree of
usage, I enjoy a balanced life, blending family
and work and leaving ample amounts of
leisure to write, play music, and visit
relatives, and enough disposable income to
dine out periodically and take my wife on the
town.

Compared to the world's silent majority,
I am notably well off, even pampered. It is
surely not I who am radical or extreme in my
practices. It is the Americans around me.
One of the proofs of my moderation is the fact
that I have taken the way I live and begun to
practice it in the heart of the city.

En route to life with less technology, my
family's idiosyncratic path has led us, indeed,
from the country to a small town, and from
there to St. Louis. Over a twelve-year period
we have tried and tested the life in each place
and grew enamored of its particular charm.
We could have remained anywhere. In the end,
however, we discovered that a suitable city can
contain elements of the country and the small
town, and add elements found in neither.

I speak largely from my experience of
St. Louis. To a passerby this has-been river
port may seem an odd choice. A drive on ▶

Mary Brende

66 It is surely not
I who am radical
or extreme in my
practices. It is
the Americans
around me. 99

Amish in St. Louis *(continued)*

Interstate 70 reveals the typical shapeless, formless, centerless metropolis ruled by cars and highway engineers, and peopled by anonymous commuters scurrying to work in order, presumably, to pay for the cost of their transportation and tract housing.

But there is another story, a scene invisible to the typical freeway traveler. The city proper was built long ago on a pedestrian scale and, in its heyday, boasted a population density similar to that of San Francisco. Several magnificent city parks—including Forest Park, the largest urban park in the United States and one-time stage for the World's Fair —lie within its borders, as well as numerous distinct neighborhoods that function like small towns within the city. The more we learned about St. Louis, the more we drooled over the prospect of small-town cohesion with access to big-city amenities, like a great orchestra and a cathedral with the largest mosaic in the world. And all at a cost of living eighty percent of the national average in a climate with four full seasons.

Inner St. Louis is a poor man's Boston, or a mid-westerner's New York. At the same time, its pace is slower than the coastal trade centers, languid and lazy (except on a few easy-to-avoid car corridors), like the meandering Mississippi.

Of the various neighborhoods that tempted us, we focused on Lafayette Square. Perched on a rise overlooking downtown and the Gateway Arch, this mile-square enclave was, ironically, once a suburb. In the early 1800s, stately French Victorian town houses began to rise from what had been public commons, and a beautifully landscaped park supplanted

66 The more we learned about St. Louis, the more we drooled over the prospect of small-town cohesion with access to big-city amenities. 99

a pasture. Later, at century's end, Lafayette Square was superseded, in turn, by still-more-opulent subdivisions opening up farther west, and fell into steep decline. By the 1970s a three-story, six thousand-square-foot architectural wonder, in passable condition, could be purchased for under $10,000. Urban pioneers began to move in.

Today the neighborhood is almost fully reclaimed and restored—and trendy. We could only afford to move to its fringe, a newly gentrifying section named McKinley Heights, and so we did. Now we savor a sort of village-life-in-the-city. We can walk or bike to the grocery, hardware, and drug stores, post office, library, church, and numerous family-owned restaurants (restaurants are indispensable centers of vitality and artistry in the city, and we gladly support them). Sunday afternoons will find us biking on meandering trails through a park bigger than the entire town we used to live in, or sitting reading the paper while the kids scamper by the duck pond.

Our house (a mere ninety-six years old and 2,600 square feet in size) serves not only as a residence, but a homestead. We garden in our backyard. We make soap in the kitchen and sell it at the nearby Soulard Farmer's Market, a traditional small merchants' pavilion continuously running since the 1830s. Since our home came with a two-story carriage house, I store our rickshaw in it and pedal the two miles to pick up fares downtown.

Many residents of the neighborhood, like us, shun suburban anonymity and, because of the constant demands of owning large, old houses, have become physically handy and adept. We know several carpenters, a tuckpointer, a flooring expert, and some who can practice more than one trade. As we ▶

66 Now we savor a sort of village-life-in-the-city. 99

> 66 Many of the denizens are artists or gardeners or, because they bought early or cheaply, possess the same homesteading spirit we do. 99

restore our own house (a far more important form of recycling, by the way, than saving old newspapers), our sense of kinship with them grows, and we begin to share advice and favors. Many of the denizens are artists or gardeners or, because they bought early or cheaply, possess the same homesteading spirit we do.

To praise our niche in the city is not to deny its drawbacks. St. Louis remains a city with city noises (a siren blares as I write this) and city hoodlums. And, thanks to the freeways to the west and south, our neighborhood also sometimes lies enshrouded in a haze of car exhaust. My youngest, Evan, occasionally goes into long coughing spells because of the fine soot in the air. Which brings me to the point: I can see why not everyone would want to live here.

I will freely admit that if the city offers things that rural communities lack, rural communities provide certain qualities not easily found in the city, like peace and quiet and safety. Is there a way to decide where to live—without going to the trouble of moving around as we have? It depends on the value one places on these qualities. And from what I know so far, I suspect that answering the question also involves taking a sober look at one's own skills and personal assets. If the shoe fits, step in.

Having experimented at length in both a rural area and a small town, I would urge special caution to anyone considering making a leap from the city to either sort of place. It is unlikely that many will find a ready-made Minimite community waiting with open arms to usher them through all the stages of self-sufficiency. Nor do I recommend searching out such communities as a kind of jobs/works agency. And without this help—unless you bring your own handyman skills with you— the hundreds of little tricks and pieces of know- how that rural people pick up unconsciously over a lifetime will take you a lifetime to learn. The skills can definitely be acquired; the question is whether the learning curve, for the average person, is too steep or long.

Perhaps more important, the lives of those who have grown up in rural settlements are so interwoven that, to outsiders, they will come across as closed and cliquish. Small towns haven't changed in this respect or others since Sinclair Lewis wrote *Main Street*. If you practice an exotic religion or hope to hold discourse upon Wittgenstein, you will feel like a native Swahili-speaker in Bulgaria.

For all that, the era of rural and small-town life is rapidly passing. Because of the spread of cars, Internet lines, and cell phones—and Wal-Marts and subdivisions—the city is sprawling into the remotest reaches of the country. The piece of land we purchased and later sold near the Minimites now has a subdivision next to it under way—albeit twenty-five miles from the nearest sizeable town! Be prepared for something worse than cliquishness: when you get there, there may be no there *there*.

Oddly enough, as people spread ever outwards, portions of urban areas first emptied out, then began to recover their cohesiveness. The turnover has led to a strange transposition of identities. Today, for us, "country life" may be more viable in St. Louis than it was in the small town we just left. The urban core is the only place sprawl cannot touch. City life, with its cars and callousness, seems to be migrating outward—even as ▶

> **The urban core is the only place sprawl cannot touch.**

Amish in St. Louis *(continued)*

simplicity and community regain force in the city.

Given this unusual turn, becoming better *off* may not mean packing up and relocating to Montaña. The steps to simple living may actually be rather simple. The first may consist merely in recognizing, and adapting to, features near where one already lives. And that is not so radical after all.

Living without so much technology may be less extreme than you think. It certainly is a lot easier. ∽

" Living without so much technology may be less extreme than you think. It certainly is a lot easier. "

Eric Brende's
Practical Tips for a Leaner and More Leisurely Life in a World of Technology

Guiding Insight: All too often our everyday "laborsaving" devices don't get rid of labor; they merely separate it out in time and space and create two, three, or more tasks out of what might have been one. The trick is to combine physical, social, and mental pursuits into a single, integrated experience. Technology should not unravel, but, if anything, enhance this integration.

1. *Transportation.* It is obvious that walking or biking provide not only more exercise than a car but also more social opportunities since, if you pass someone on the street, you can say hi or stop and chat. What is less obvious is that, for example, a bicycle makes a wonderful cart for practical errands—with the right attachments. We use pairs of collapsible rear bicycle baskets each holding a full sack of groceries or two gallons of milk. We've seen others pulling small trailers. No less important for our daily transit are the Trail-A-Bike and the Burley Piccolo, devices that transform an adult bicycle into a tandem for a child. Finally, I ride a recumbent—a Sun EZ-1 Super Cruizer —allowing me to recline on a comfortable seat, save my back, and get more power with each stroke. (It's heavy, so if you want a more maneuverable model, get the aluminum frame.) Admittedly, the life of practical ▶

9

errand running by bicycle would not be easy in suburbs with dendritic street systems feeding into impassable car arteries. If such is your fate, consider relocating further citywards.

2. *Entertainment.* Obviously it is more wholesome to read your children a story than sit together mesmerized by television or a video of *Star Wars.* But there is another alternative we've found as well—listening to radio or books on tape, preferably as we eat or work together on a family project. Unlike TV, audio entertainment or educational material can readily blend into a larger experience without overwhelming it. But it goes without saying that live events surpass simulated or electronically transmitted ones. St. Louis is a mecca for families because so many of the major attractions are free—the zoo, a nature preserve known as Grant's Farm, the summer Shakespeare festival, sections of the outdoor musical theatre (the MUNY), and, on a limited basis, the stellar St. Louis Symphony.

3. *Communication.* Using the computer at the library, but not at home, does two things for me. First, it gives me an excuse to get out and ride my bike. Second, it limits my time to a manageable segment. Having been requested to use e-mail for the convenience of my publisher, I began to attract all kinds of cyber-missives. Since I don't want to encourage these, and since I now use the Internet only two or three times a week, I have deployed an automatic message politely informing callers of my home mailing address and phone number.

> 66 It goes without saying that live events surpass simulated or electronically transmitted ones. 99

4. *Washing and cleaning.* Instead of taking an aerobics class, get a swing-handled washing machine, and make your laundry your workout. A load takes two hundred strokes and two rinse basins. We don't recommend using a wringer because it breaks buttons and is relatively laborious. Instead, give each item a quick twist (good for the gripping muscles) and let it drip dry. In a pinch, if you own an electric washer, you can put the rinsed clothes through the spin cycle to shed the excess water before hanging. And need we mention that a Bissell carpet sweeper is much lighter and easier to use than a power vacuum? It works best for a light touch-up. If you must suck up all the dirt from a rug, then use the vac.

5. *Tools and lawn implements.* I rarely power drill or power saw, though neighbors have given or loaned me these items. If the point is to get it over with quickly, then the work is not enjoyed for its own sake, something that is made much easier when things are quieter and more slow-paced. Still, a hand drill can often be as fast as the power substitutes. I finally did get a "Yankee Drill" from a reader who saw my interest in it and wanted to trade one for a copy of my master's thesis. The thing is whiz-bang amazing and can fit in tight places, which a crank drill cannot. You can special-order one from Stanley Tools. If you really want to know what to do with a saw and hammer, I recommend taking a summer homebuilding course, as I did, from the Shelter Institute in Bath, Maine. We love our hand-powered lawn mower; these are sold now at the big home improvement stores, and probably hardware stores, too. The one trick is to ▶

> 66 If the point is to get it over with quickly, then the work is not enjoyed for its own sake, something that is made much easier when things are quieter and more slow-paced. 99

adjust the stationary blade every year or two using the pair of screws at either end. This tightens up the mower's scissor action. Speaking of scissors, we have shunned the noisy weed-eater in favor of a pair of long-handled shears, which we ordered from Gardener's Supply catalog.

6. *Food.* We are what we eat. Our community, also, is shaped by what we eat. We've filled up our small backyard with plantings, most of them edible, and treat it as an extension of the kitchen. What we don't grow we purchase from the farmer's market and a small neighborhood grocery. We enjoy dining out at least once a week, almost always at family-owned or ethnic restaurants. Food preparation is one of the last havens from all-out mechanization because no machine or corporate franchise can yet duplicate the finesse of a skilled chef within the time constraints that hot and fresh food requires. We want to promote this kind of skill, immediacy, and small proprietorship.

7. *Finances.* If you've cut back this much on technology, you may be able to afford more than occasional dining out. You may be able to pay off your mortgage faster. Doing that, in turn, will ultimately free you to live more simply outside the mechanical system. We, however, did the reverse: worked hard for two years so we could pay for our property first, *then* lived more freely without much technology.

> " Food preparation is one of the last havens from all-out mechanization. "

8. *Household management.* Too much integration is not a good thing. When both husband and wife are working on the same premises and don't wish to get entangled in each other's hair, it is imperative to maintain clear lines of authority and workspace. I was handicapped in our previous house until I had my own office for writing—well out of earshot from the rest of the family. Mary, in turn, was stymied when my writing materials took over the kitchen island, which doubled as her homeschooling podium, or when my need for quiet quashed the children's need to play. Now in our new St. Louis abode, I look forward to converting a section of the basement with generous south-facing windows into my office getaway. The basement room will also be good for cutting and curing soap, and an adjoining area will serve for woodworking. There, too, children can have workspaces for their own projects. Good walls make good working households.

9. *Community.* One thing we love about the city is the availability of so many levels of human relationships, from casual to intimate, crass to reverent. There is also a motley selection of clubs and associations, many of which pertain to varying degrees to my low-tech interests. As a newcomer I have only begun to take part in a homeschooling cooperative, scouts, and our church. Had I the time or inclination, I would look more deeply into the bicycle federation, the small business association, and the neighborhood betterment group. ▶

> " When both husband and wife are working on the same premises and don't wish to get entangled in each other's hair, it is imperative to maintain clear lines of authority and workspace. "

10.

Reflection. Meditate before you act. Our modern technological world has ultimately grown out of a certain way of thinking, a compulsion to reduce reality to simple, easy-to-manipulate components, and thereby control it. To this mistrust might be opposed the Anabaptist term *Gelassenheit,* or "self-surrender": the acknowledgment that not everything can, or should, be controlled, and that the deeper mysteries of life must be savored and pondered. With *Gelassenheit,* any changes in technology will be more deliberate and circumspect, not sudden and impulsive. The Minimites allow themselves an experimental period, testing the implications of this or that device, then evaluating it and making further adjustments. Their openness to observation and willingness to adapt— and humility before the complex tangle that constitutes Truth—should be a lesson for the rest of us. Practical action will come to naught without reflection. ᕦ

" Meditate before you act. "

14

Recommended **Reading**

John Hostetler. *Amish Society* **(4th ed.).
Johns Hopkins University Press: 1993.**

Provides an excellent historical and sociological
background of Old Order Amish and
Mennonite groups, in all their interesting
variety. Hostetler was raised Amish, and
provides both a sympathetic and critical
understanding. A must for anyone wanting
to learn more about these little-understood
peoples.

Helen and Scott Nearing. *The Good Life*
(contains *Living the Good Life* **and**
*Continuing the Good Life***). Shocken: 1990.**

The classic accounts of successful rural
simplicity and self-sufficiency, engagingly
and rigorously presented. The Nearings'
insights about motorized technology and
horsepower are particularly penetrating.
Note, however, that this pioneering couple
undertook their venture (in the 1930s) more
to make a political, than technological,
statement—namely by showing how to live
well without being dependent on the corrupt
power structures of industrial capitalism.

James Howard Kunstler. *Home from
Nowhere: Remaking Our Everyday World for
the 21st Century.* **Simon & Schuster: 1996.**

A rollicking account of what makes an urban
neighborhood truly livable, workable, and
walkable, drawing chiefly on the discoveries
of the New Urbanist movement led by Andres
Duany (whose book *Suburban Nation* is ▶

also fascinating, and takes up where Kunstler's leaves off). The insights in these books will dramatically clarify anyone's discernment of the desirable attributes of an urban setting and what it takes to recover them.

**Katie Alvord. *Divorce Your Car.*
New Society Publishers: Canada, 2000.**

If the most counterproductive machine ever made was the car, the most comprehensive book ever written about unmaking it was this one. Alvord provides a grim history of the automobile's rise coupled with practical pointers aiding the average person's recovery from it.

**E. M. Forster. *The Machine Stops.*
(Originally published in *The Eternal Moment and Other Stories*). Harcourt, Brace, and Jovanovich: 1928.**

For someone who lived before computers and television, Forster was uncannily prescient. This poetic novella describes our present personalized audiovisual cocoon eight decades before it surfaced, and explores its grim physical and social consequences. ◁▷